The Scarlet Lancers

General John Burgoyne, "Gentleman Johnny"; soldier, playwright and Member of Parliament, raised 16th Light Dragoons in 1759. Portrait by Sir Joshua Reynolds. *The Frick Collection*

THE SCARLET LANCERS

The Story of 16th/5th The Queen's Royal Lancers

1689 - 1992

James Lunt

Leo Cooper
LONDON

First published in Great Britain in 1993
by Leo Cooper, 190 Shaftesbury Avenue, London WC2H 8JL
an imprint of Pen & Sword Books Ltd., 47 Church Street,
Barnsley, S. Yorks S70 2AS

A CIP catalogue record for this book is available
from the British Library

ISBN: 0 85052 321 4

Printed and bound by The Bath Press Ltd., Avon

TO THE REGIMENT

Contents

FOREWORD
by
THE COLONEL OF THE REGIMENT

This is not a conventional Regimental history. It does not attempt to provide a continuous and detailed account of activities over the years — we already have volumes doing that for the 5th Lancers, the 16th Lancers and the 16th/5th Lancers from their respective formations up to 1961. Instead James Lunt (both a well-known author and a senior ex-officer of the 16th/5th Lancers) uses his own inimitable style to describe specific events and individuals and uses these to illustrate why these distinguished Regiments developed unique personalities.

He deals rather more fully with the recent activities of the 16th/5th Lancers, particularly those from 1964 to 1991, not least because the Regiment was heavily involved in a remarkably high proportion of the active service engagements of the British Army in this period. Aden and the Radfan, two years in Northern Ireland at the height of the troubles, Cyprus during the Turkish invasion, Beirut and the Gulf War are all covered, providing for the first time a coherent and fascinating view of the Regiment's so called "peacetime soldiering".

The 16th/5th Lancers is to amalgamate with the 17th/21st Lancers in the summer of 1993 to form a new Regiment to be titled "The Queen's Royal Lancers". Some might argue that James Lunt's book chronicles the end of the 16th/5th Lancers. I disagree. Rather it marks the end of one chapter in a book that has many chapters yet to be written. James Lunt uses history to demonstrate that we may confidently expect The Queen's Royal Lancers to uphold and indeed build on the distinguished records of its predecessors.

Baughurst, October 1991   *Alastair Dennis*

INTRODUCTION

THIS is the story of two cavalry regiments of the British Army, one raised in 1689 and the other in 1759, that came together in 1922 to form a third regiment, the 16th/5th The Queen's Royal Lancers. Since the older of the two regiments of this union, the 5th Royal Irish Lancers, had had a somewhat chequered career, having been disbanded in 1799 and re-formed in 1858, it ranked junior to the 16th The Queen's Lancers. This accounts for the 16th preceding the 5th in the title after amalgamation, a source of some perplexity for those who feel it should be the other way round. During the fighting in Italy in the Second World War, the Americans used to refer to British cavalry regiments with this kind of numbering as 'fractionalized' regiments.

As any reader of the Marquess of Anglesey's magisterial *History of the British Cavalry* will be aware, cavalry regiments of the line in the British Army have been subjected to frequent change, sometimes disbanded, occasionally re-formed, and more recently often amalgamated with another cavalry regiment. When I set out to write this book in September, 1988, any one who had predicted the imminent breakup of the Soviet Empire, the demolition of the Berlin Wall and the unification of the two Germanies, and a war against Iraq under the leadership of the United States, would almost certainly have been laughed to scorn.

However, this is what happened, changing the map of Europe, and having moreover a dramatic effect on the future shape and size of the British Army. With the withdrawal of the Soviet occupation forces from Eastern Europe, it has made possible a considerable reduction in strength of the British Army of the Rhine (BAOR). This, coupled with reductions elsewhere, will mean a cut in the overall strength of the British Army of at least 50,000 officers and soldiers. Needless to say this must mean the disbandment or amalgamation of many distinguished cavalry and infantry regiments with their roots buried deep in Britain's past. Not the least among them must be the 16th/5th The Queen's Royal Lancers because the Royal Armoured Corps is to bear a high percentage of the forthcoming reductions. Therefore it might be argued by those so inclined that what follows in these pages is more an obituary than an historical narrative, but I believe that to be an unnecessarily gloomy view. Those who took this line in 1922 when the 16th and 5th Lancers were amalgamated, and there were more than a few of them, were later shown to be wrong. The regiment that resulted from the union was more than worthy of its illustrious predecessors.

It is of course permissible to question the Government's wisdom in making such large reductions in the Army's fighting strength. As we have seen in the case of the recent campaigns in the Falklands and Iraq, wars have a habit of breaking out when and where they are least expected. In my judgement it would have been wiser to have cut more deeply into the 'tail' in order to preserve the 'teeth'. The Ministry of Defence is certainly crying out to be reduced by at least fifty percent; too many admirals, generals and air marshals, together with their civil service counterparts, looking after too few warships, soldiers and planes. But perhaps this is too much to be hoped for.

Whatever the future may hold in the way of amalgamations, it is essential that we should retain the regimental system which has proved to be the bedrock of the British Army. It has been shown time and again that the merging of two regiments with proud traditions will strengthen rather than weaken the new regiment that will result from their union. There is more than enough evidence to prove this point. It was certainly true of the

16th/5th Lancers during the hard-fought campaigns in North Africa and Italy in the Second World War, and again more recently in Iraq. I am certain it will be repeated in the future. Although naturally I shall regret the passing from the *Army List* of the Regiment in which I, and my son after me, had the privilege of serving, I am sure that the phoenix which will arise from our ashes will preserve all our traditions and in the course of time create its own.

Tradition, as we know it in the British Army, is hard to define. It has only the most slender connection with the 'Tradition' recently depicted in a BBC Television Series of that name. That seemed to consist largely of officers in mess-kit lounging round mess tables laden with silver, and passing the port to each other. This may be the image that an ignorant public, and an equally ignorant television producer, has of the Army, and it is perhaps the most photogenic aspect, but it is as far removed from the real meaning as is the tank from the horse, or the sword and lance from the Swingfire guided missile. Tradition, I would suggest, is more of the spirit than anything more tangible. A regiment's deeds in the past help to strengthen it today, and in the future. As an example of what I mean, it was the custom on band nights in the officers' mess of the King's Own Yorkshire Light Infantry for the Vice-President after the Loyal Toast to give the toast, 'Ensign Dyas and the Stormers'. This commemorated the outstanding gallantry of Ensign Dyas and the 'Forlorn Hope' of the 51st Light Infantry which he led into the breach at the storming of Badajoz in Spain in 1811. I find it comforting that such bravery should be recalled nearly two centuries later. This is what Tradition is all about.

There are of course many, both in the Army and outside it, who regard the perpetuation of the regimental system as an obstacle to administrative reform. They would like to see a Corps of Infantry and a Royal Armoured Corps in which both officers and soldiers can be freely cross-posted between the units that make up the Corps. This has been the case within the Royal Regiment of Artillery since the abolition of the separate branches of Horse, Field and Garrison after the First World War, and is similar to the practice in both the French and US

Armies. There is, however, much more cross-posting between units nowadays than was the case between the two world wars; it was certainly common enough in the days when officers purchased their commissions and' subsequent steps in rank, abolished by Cardwell in 1871. The Duke of Wellington served in no less than six regiments in the course of his progress up the military ladder. I myself served in two.

Whenever I find myself called upon to defend the regimental system, I am reminded of F. Majdalany's account of the Battle of Cassino in Italy in 1944. Majdalany was serving at the time in the Royal Sussex Regiment:

'Under-supplied, without sufficient time to prepare, these few fought a lonely battle in the mountains and no one in the rest of the army had any idea what they were fighting. They had nothing to sustain them except that potent imponderable their regimental identity. It mattered to the Rajputana Rifles that they were Rajputana Rifles. It mattered to the Royal Sussex that they were Royal Sussex. In the end it was probably this alone that enabled them to hang on.'

I believe this says it all. And I leave it to the reader to judge whether we in the 16th/5th The Queen's Royal Lancers were faithful to the traditions we inherited, and which we pass on to our successors in the certain hope that they will be equally cherished in the future.

Oxford, June 1991.   *James Lunt.*

CHAPTER ONE

The Royal Dragoons of Ireland

THE English, with their concern for civil liberty, have always been suspicious of a Standing Army, even before Oliver Cromwell's military rule convinced them that they were right.* But in 1689 Parliament legalized the existence of a Standing Army by passing the Mutiny Act, the intention of which was to prevent desertion from the army of King William III to the army of King James II. It has, however, been usually regarded as the beginning of the Standing Army, although in fact an army of sorts had been in existence for many years previously. When regiments of Foot or Horse had to be raised in those days, it was the custom for the King, or some official acting on his behalf, to call upon gentlemen of substance in the country to raise so many officers and men, the regiment so formed usually taking for its title the name of the man who had raised it. Thereafter to all intents and purposes the regiment became the property of its Colonel who might not in fact lead it in battle − he could appoint a Lieutenant-Colonel to do that for him − but who would confidently expect to make a lot of money out of his investment. In 1689 such a regiment of Horse was raised in Enniskillen in Northern Ireland on the orders of Gustavus Hamilton, Governor of Enniskillen. The man entrusted with this commission was Captain James Wynne, 'a gentleman of Ireland, but then a captain in Colonel Stuart's Regiment'. Wynne was appointed a Colonel of Dragoons with the pay of 15 shillings a day. His regiment's title was 'Wynne's Inniskilling Dragoons'.

---

* I use the word 'English' advisedly. The Scots have always been much more conscious of their regiments than the English.

Wynne's Inniskilling Dragoons 1689. A silver statuette

It is a melancholy reminder of the state of Anglo-Irish relations down the centuries that three hundred years ago Protestants and Catholics stood face to face, as they do today, at the gates of Londonderry. King James II, recently ousted from his throne, had landed in Ireland and placed his fate in the hands of the Earl of Tyrconnel. Protestants in the north rallied under Gustavus Hamilton to defend the cause of King William III. Two regiments of Dragoons and three of Foot were hastily raised in Enniskillen. In July, 1689, General Kirk arrived from England to relieve Londonderry, and on the 20th of that month issued commissions to the officers of the Enniskillen regiments in the name of King William III, who subsequently ratified them. Together with the other Enniskillen regiments, or more

The Duke of Marlborough campaigning in the Low Countries. Ross's Dragoons were "in the thick of the fight" at Blenheim. An oil painting by R. Hillingford

The Honourable Hugh Somerville joined the 16th Light Dragoons on its formation in 1759. Lieutenant-Colonel 1763. Portrait by Tilly Kettle

properly levies, Wynne's regiment was brought on the establishment of the Regular (or Standing) Army by a Royal Warrant dated 'the first day of January, 1689, in the first year of our Reign'.

It was not an efficient unit, even when judged by the not very exacting standards of the time, and has been compared to a mob of undisciplined boys led by officers who were ignorant, negligent and drunken. However, they could fight well enough, drunk or sober. In those days Dragoons were essentially mounted infantry, using their horses to convey them to a suitable tactical position, dismounting thereafter and fighting on foot. More of Wynne's regiment probably went on foot than in the saddle, horses being in short supply in Ireland. So bad were the arrangements for the shipment of horses from England to Ireland that one regiment lost every horse during the sea passage. It was fortunate for William III that Tyrconnel's troops were in no better shape, and that he was given time to get the raw Enniskillen levies into some kind of order.

This was the task of Marshal Schomberg, a German veteran reputedly aged 80, who had been appointed Commander-in-Chief by King William. He arrived in August, 1689. By then Wynne's regiment had been blooded at the Battle of Newton Butler towards the end of July, when we are told 'the name of the Enniskillen men became a terror to the Irish'. Schomberg was so struck by 'the horrid and detestable crimes of profane cursing, swearing, and taking God's Holy Name in vain,' that he issued a special order of the day on the subject, dated 18 January, 1690, in which he charged and commanded all officers and soldiers to 'forbear all vain cursing, etc etc'. It would seem that in this respect at least soldiers have not changed all that much down the centuries.

William III took the field himself on 14 June, 1690, and shortly afterwards defeated James II at the Battle of the Boyne on 1 July. The battle started inauspiciously for King William. His Commander-in-Chief, Schomberg, was killed at the outset while reconnoitring the river bank. Tradition has it that he was shot by an Irishman with his duck gun. He was buried in St Patrick's Cathedral in Dublin where more than 200 years later the 5th Royal Irish Lancers erected a monument to commemorate their comrades who had fallen in the South African War; the monument is close beside Schomberg's grave.

The Battle of the Boyne was hard fought and at one stage William found it necessary to place himself at the head of the Dragoons from Enniskillen, saying, 'Gentlemen, I have heard much of your exploits, and now I shall witness them'. The English cavalry then crossed the river and charged with such fury that they scattered the enemy infantry but then galloped off out of control. Fortunately for King William the Irish infantry broke at the crucial moment and took to their heels; they were only saved by the gallantry of the Irish cavalry who charged time after time. The pursuit after the battle was nevertheless a bloody one.

Wynne's Dragoons later took part in the unsuccessful sieges of Limerick and Athlone. There is a sadly contemporary note in the description of an assault at Limerick during August, 1690: 'For three hours did a sharp fight continue, in which the Irish women boldly joined; and when they failed to obtain more deadly missiles, threw stones and broken bottles' (shades of the Falls Road, 1970!). Wynne's Dragoons were also present at the Battle of Aughrim on 12 July, 1691, which virtually concluded the war and where, according to an eyewitness, Captain Parker, the Irish never fought so well in their own country as they did on that day, although their gallantry was in vain.

In May, 1694, Wynne's Dragoons were sent to Flanders to join the allied armies collecting there to fight the French. Wynne's regiment was ill-horsed, most of the animals out of condition after a long time at sea while crossing the Irish Sea and English Channel. There was, however, time enough to improve on this since warfare at that time can hardly be described as fast-moving. It was more a matter of sieges and patrol actions than set-piece battles; when the latter did take place, it was all according to the manuals of war and extremely formalized. But men were killed, and men were wounded, most of the latter dying, since medical attention was mainly confined to amputations and to bleeding. Wynne was wounded in the knee at Moorsleede, where his Dragoons defended a convoy

he Battle of the Boyne

of supplies against enemy attack. Although the wound was slight, Wynne died from it. King William gave the colonelcy to Charles Ross who had served as one of his aides-de-camp. In July, 1695, Wynne's became Ross's Dragoons.

The war in the Low Countries ended in 1698. Ross's Dragoons lost far more men from disease than from enemy action and like every other regiment they had many deserters. The wonder of it is that men could be induced to serve under such wretched conditions, with inefficient officers, scoundrelly commissaries, and doctors who made better butchers than surgeons. But for men who came from the backward rural areas of Ireland, life on campaign in Flanders was in some ways preferable to the brutish conditions at home; at least there were far better opportunities for loot. Ross's Dragoons returned to Ireland where the strength of the garrison was fixed at 12,000 men. Ross's Dragoons were reduced to eight troops, amounting in all to 362 officers and soldiers, who were billeted in inns and lodging-houses throughout the Province of Connaught.

Maintaining discipline among these scattered detachments was always a problem in Ireland, where the army operated as a gendarmerie, and although life was on the whole easy-going, punishments when ordered could be draconian. Six men from Ross's Dragoons were sentenced by Court Martial in December, 1698, 'to run and be whipped several times by an entire regiment of Foot drawn out for that purpose on three several days on St Stephen's Green'. The punishment, known as the Gatloup, was carried out by troops paraded in open ranks. Each man carried a stout stick. The ranks were faced inwards and the prisoner, stripped to the waist, was marched up and down the lines of men. Each man was expected to strike him with maximum force on the 'naked back, breast, arms, or where his cudgel should light,' while the screams of the victims were intended to be drowned by the drums which beat throughout the punishment. Many men died while undergoing this savage ordeal.

In March, 1702, Ross's Dragoons were dispatched to join Marlborough's army in the Low Countries. They were present at Blenheim where they were in the thick of the fight. Marlborough's pencilled note to his wife, Sarah, written in the fading light after a long and exhausting day, must surely be one of the shortest dispatches in history:

'13 August, 1704

I have not time to say more, but to beg you will give my duty to the Queen, and let her know her army has had a glorious victory. Monsr. Tallard and two other generals are in my coach, and I am following the rest. The bearer, my aide-de-camp, Colonel Parke, will give her an account of what has pass'd. I shall do it in a day or two by another more at large.'

Ross had petitioned that his Regiment should be known as The Royal Dragoons of Ireland and in March, 1704, that title was formally conferred on the Regiment. Therefore it was as the

Royal Dragoons of Ireland that the Regiment charged at Blenheim, capturing three French kettle-drums which Marlborough directed should henceforward be carried at the head of the Regiment. These kettle-drums are reputed to be those still in existence in the Queen's Armoury in the Tower of London. When the Regiment received its first Guidon from the Queen at Buckingham Palace on 19 March, 1959, the Blenheim kettle-drums were brought from the Tower and piled in front of the parade. On them was laid the Guidon before its consecration by the Chaplain General to the Forces, 255 years after their capture at Blenheim.

In 1706 the Royal Dragoons of Ireland, brigaded with the Scots Dragoons (Scots Greys), distinguished themselves at the Battle of Ramillies on 23 May. The two regiments captured the entire King's Regiment (*Régiment du Roi*) while charging the enemy, and followed this by taking two battalions of the *Régiment de Picardie* prisoner, and destroying the third battalion. For this feat both regiments of Dragoons, Scots and Irish, were given the distinction of wearing Grenadier caps, thereby differentiating them from the rest of the cavalry. Captain Molesworth of the Regiment, who was serving as aide-de-camp to Marlborough, was instrumental in saving his General's life. The Duke, in the thick of the fighting, was thrown from his horse and was in imminent danger of capture. Dismounting, Molesworth handed his horse to the Duke, who was able to escape, leaving Molesworth on his own. He managed to recover the Duke's horse and took it back to Marlborough. While the Duke was remounting, the equerry holding his stirrup had his head taken off by a round shot. Molesworth later had a medal struck in his honour, a very unusual distinction. He reached the rank of General, as Viscount Molesworth, and was Colonel of the Royal Dragoons of Ireland from 1737 to 1758.

Ramillies was probably Marlborough's hardest fought battle; he was in the saddle for nearly 24 hours, and, as already described, was in the thick of the fighting at times. The Royal Dragoons of Ireland were also present at Marlborough's other two great victories, Oudenarde on 11 July, 1708, and

The 5th Royal Irish Dragoons about 1740

Malplaquet on 11 September, 1709. They helped to capture Bruges in 1706 and took part in what many regard as the most brilliant of Marlborough's operations — the passage of the French lines on 4 August, 1711. It is sad that this brilliant feat should have been followed by Marlborough's recall and subsequent disgrace. Few British commanders have been loved so well by their soldiers as Marlborough, or 'Corporal John', as they called him.*

Life in the ranks was still as hard and as brutish as it had been when campaigning under William III, but Marlborough did at least give them victories and organized a more reliable supply system. There was a great deal of peculation and nepotism, as is evidenced by this letter from Marlborough to Lord Cardigan,

* General 'Bill' Slim comes nearest to Marlborough in receiving the affection of his soldiers, in my opinion.

who had requested that the son of the late Major-General Brudenell should be given a company:

'I have so just a sense of the father's good services that I shall always be glad to embrace any opportunity of showing it to his family; but your Lordship tells me he is not above five years old!'

Charles Ross, now a Lieutenant-General of Horse, took some part in the negotiations which concluded the war. In June, 1713, he was told 'the Regiment under your command is to be put on the Establishment of Ireland and to be paid for by the revenues of that country'. Ross himself was sent to Paris three months later as Envoy Extraordinary. He was lucky to avoid the slow deterioration in military efficiency that seemed to be inseparable from service in Ireland during the 18th century.

It was to be the fate of Ross's Regiment to serve continuously in Ireland for the next 86 years. The 5th Royal Irish Dragoons, as they are shown in the 1752 Army List, were never given the opportunity to escape from the dreary round of garrison duties that was the lot of regiments on the Irish establishment. They hunted down smugglers, galloped after highwaymen, held down the discontented and impoverished peasantry, and came together only once or twice annually to take part in a review. Writing of the cavalry regiments on the Irish establishment, Sir John Fortescue says they were 'absolutely useless and untrustworthy'.* He also says they were 'dispersed in small parties all over the country for the protection of isolated buildings and individuals. This in itself was sufficient to ruin all discipline; and the evil was not mitigated by the absence of great numbers of officers from their posts'.**

Not everyone was critical. 'Cavalry Corps in Ireland were extremely select,' wrote Surgeon John Smet of the 8th Light Dragoons in 1784, 'as from the very low establishment, it was in the power of the Colonels of choosing among a number of young gentlemen of distinction who might wish to get a commission, and who all could easily afford to add a hundred pounds a year to their pay. The warrants were also purchased at a high price, often by the sons of gentlemen for as much as five hundred guineas. The privates were always young men well recommended and whose connections were known.

* *History of the British Army* Vol 3, p.532
** *Ibid.*, Vol 4, p.518

Indeed, the dragoon service was at that time extremely easy and pleasant, so much so, that when a vacancy happened, several desirable recruits always offered, and the men selected in general, got no more than one shilling bounty'.*

Pleasant it may have been, but it was certainly not soldiering. Smet tells us that two-thirds of the officers were away for most of the year on leave. The men did not do too badly either. They took their horses home with them and seldom wore uniform. To save the expense of forage, horses were put out to grass for as long as possible and must in consequence have been unfit for hard work for most of the year. 'Such a service had many attractions,' says Smet. The annual review brought everyone together again, and, 'The Officers now meeting again, after such a long separation from each other, in affluent circumstances, which they had improved while they lived with their friends, justly looked on the time of year they were to be reviewed in as the pleasantest season. The mornings were spent at exercise and the remainder of the time in festivity'.

It was not always easy to find a field of suitable size on which to review a regiment. Lieutenant-Colonel the Hon Charles Stewart, brother of the Foreign Secretary, Lord Castlereagh, was commanding the 5th Royal Irish Dragoons in 1785 when he drew attention to the problem in a letter from Clonmel on 16 June:

'That your Memorialist took a field of exercise for the 5th Dragoons during their assembly and review at Cashell, in the last and present months, as there was no common ground near that town. That he was obliged to pay 13 guineas for that field, as he could not get a proper place for less money. Praying to be allowed to charge the said sum to the contingent bill of said Regiment in the usual manner.'

We are not told whether the Commanding Officer was successful or not.

Charles Stewart presumably had purchased command of the 5th Royal Irish Dragoons when he was still very young. He could not have been better connected, son of the Marquess of Londonderry and brother of the Foreign Secretary. He was Wellington's Adjutant-General in Spain and acquired a certain

---

* *Historical Record of the 8th Hussars* (Smet)

reputation as a cavalry leader, although it was said that he was hampered in the field by his poor eyesight. He became Marquess of Londonderry in due course, as well as a General. Stewart was specifically absolved by Cornwallis (Lord-Lieutenant in Dublin 1798-1801) of any responsibility for the poor state of the 5th Royal Irish Dragoons when it was decided by George III to disband the Regiment; Stewart was at the time commanding the 18th Light Dragoons which he took to Portugal, but he cannot be entirely innocent of blame. It is a truism in the British Army that a regiment takes its tone from its commanding officer. If that were the case with the 5th Royal Irish Dragoons, Stewart would appear to have failed.

Ireland was in a sorry state towards the end of the eighteenth century. There was appalling poverty and almost continuous discontent. Had this not been the case, there would have been no requirement to scatter the troops round the countryside in small groups to keep the peasantry in order. The consequences of such service were absentee officers and bored and demoralized soldiers. Mutiny was not infrequent, insub-ordination was common, and desertion an almost daily occurrence. When General Abercromby arrived in Dublin as Commander-in-Chief in 1798, he was appalled at the state of affairs he found. 'The very disgraceful frequency of courts-martial, and the many complaints of irregularities in the conduct of troops in the Kingdom, have too unfortunately proved the army to be in a state of licentiousness which must render it formidable to every one but the enemy,' stated Abercromby's first General Order. He followed it up by resigning, handing over to his deputy, General Lake, who was later to find India a more rewarding country than Ireland in which to win military distinction.

If regiments recently arrived in Ireland succumbed so quickly to the general malaise, it is hardly surprising that a regiment like the 5th Royal Irish Dragoons, which had known no other garrison since Marlborough's campaigns, should have been in such a sorry state. Their inspection report in 1797 could not have been plainer: 'A very essential change is absolutely necessary to put the 5th Regiment of Dragoons in a condition for service which at present they are entirely unfit for'. Nothing was done, however. The good officers and NCOs took themselves off, leaving behind the lazy, drink-sodden and incompetent. On 23 May, 1798, a rising took place in Ireland and the detachments of the 5th Royal Irish Dragoons found themselves fighting for their lives. Like all rebellions of its kind, that in Ireland in 1798 was marked by cruelties on both sides. There were various engagements, the best known of which is Vinegar Hill, but by September the authorities had regained the upper hand. There was now time to consider the problem of the Irish garrison in general, and of the 5th Royal Irish Dragoons in particular.

The regiments on the Irish establishment had been found deficient in the essential military virtues, but it is not clear why the 5th Royal Irish Dragoons were made the scapegoat for the failings of the rest of the garrison of Ireland. This is the more surprising in view of the fact that its last Commanding Officer had very powerful connections in the government, and not only with his brother, the Foreign Secretary. The whole business remains obscure and may owe something to the fact that King George III, already mentally unstable, was obsessed with the fear that his Prime Minister, William Pitt, was determined to bring about Catholic emancipation in Ireland. It may have been thought that by taking such drastic action as the disbandment of one of the oldest and most senior regiments on the Irish establishment, it would be clear how strongly the King was opposed to any suggestion of Catholic emancipation. Beyond this, which can only be supposition, it is hard to know why the government acted as it did.

Unfortunately there can be no shadow of doubt that the Regiment was in a very poor state, although whether this was so bad as to merit Lord Cornwallis's strictures in his letter of 1 January, 1799, to the Secretary at War (the Duke of Portland) can only be a matter of conjecture.

'The highly improper, dangerous and disloyal conduct of the non-commissioned officers and private soldiers of the 5th or Royal Irish Regiment of Dragoons, and generally speaking the irregularity, the want of discipline, and inattention of the officers, have given me much anxiety and uneasiness for some time past,' wrote Cornwallis, 'but at present, the information I have received respecting them is of so serious a nature that I am under the necessity of reporting their misconduct, formally, to your Grace for the immediate information of His Majesty'. He went on to say that some deserters from the Regiment had been apprehended and would be tried by a General Court Martial, but this, in his judgement, would not be enough. He strongly recommended the removal of the Regiment from the Irish garrison.

We do not know whether Cornwallis was surprised by the reaction in Whitehall. In a despatch dated 12 January the Secretary at War expressed the King's concern and displeasure at the conduct of the Regiment, going on to say, 'His Majesty has therefore determined that the Fifth or Royal Irish Regiment of Dragoons shall be immediately disbanded'. The Regiment was concentrated in Newry prior to embarkation under the guns of the frigate, HMS *Ariadne*, after which it was conveyed to Liverpool early in February, 1799. It then proceeded on foot by march route to Maidstone, and thereafter to Chatham where it was formally disbanded on 10 April, 1799.

Whatever their failings may have been in the past, the 5th Royal Regiment of Dragoons in their hour of trial appear to have behaved in exemplary fashion. A report on the Regiment's last days comments favourably on its good behaviour while awaiting disbandment. 'We cannot conclude,' says the report, 'without expressing our regret that a Regiment which has so frequently deserved well of its country should have incurred His Majesty's displeasure. Let us hope that, like the Phoenix, it may some time or other rise out of its own ashes, be restored to the Army, and add fresh laurels to those of Ramillies and Hochstet [Blenheim].'

As far as can be established there were only three soldiers from the 5th Royal Irish Dragoons who were court-martialled for belonging to the Fenians (United Irishmen). Two of them were executed; the other turned King's evidence and was transported for life. There is nothing to show that others in the Regiment had been disloyal. Many of the officers and soldiers were transferred to the 18th Light Dragoons under their former Commanding Officer, the Honourable Charles Stewart. There could have been no one better to separate the sheep from the goats!

The 5th Royal Irish Dragoons had been granted a special kettle-drummers' allowance by the Duke of Marlborough to commemorate their capture of the French kettle-drums at Blenheim. This amounted to six pounds, four shillings and has never been claimed since. Nor has any trace been found of the Regimental Standard which may have been claimed by the Regimental Colonel at the time, General Lord Rossmore.

The exact circumstances of the disbandment of the 5th Royal Irish Dragoons remain a mystery to this day. None of the Dragoon Guard regiments on the Irish establishment, formerly the First to the Fourth Horse, converted in 1788 to the Fourth to the Seventh Dragoon Guards, described by Fortescue as being 'valueless and obsolete,' were disbanded. Instead it was only the 5th Royal Irish Regiment of Dragoons who were, in Fortescue's words, 'swept... with ignominy from the list of the Army'.

CHAPTER TWO

Burgoyne's Light Horse

During the wars between Frederick the Great and the Austro-Hungarian Empire a new kind of cavalry appeared on the battlefields of Central Europe. They came to be known as Hussars, a Hungarian word by derivation, and they were primarily intended for outpost duties and as scouts. Recruited from the great plains of Hungary where living on horseback was a way of life, they were very lightly equipped and mounted on ponies which could live off the country and make up in speed and endurance for what they lacked in looks. It was not long before other armies began to copy the Austrians, and as a result there developed three distinct types of cavalry — heavy cavalry for shock action, light cavalry for reconnaissance, and a form of mounted infantry which most closely approximated to the cavalry of the preceding 200 years. In the latter case the horse was merely the means for transporting the soldier to the scene of action where thereafter he fought on foot; cavalry of this kind were usually known as Dragoons.

In 1759 the British Army decided to follow the example of the continental armies and form regiments of light cavalry to be styled Light Dragoons (later Hussars and Lancers). On 17 March, 1759, orders were issued for the raising of the 15th Light Dragoons, and four months later similar orders were issued for the raising of the 16th Light Dragoons. It was then the practice, whenever a new regiment was to be raised, to nominate either a country magnate or some sufficiently distinguished serving officer to form the regiment. In the case of the 16th Light Dragoons the officer nominated was John Burgoyne of the 2nd Foot Guards. He is known chiefly, if not entirely, for his surrender to superior enemy forces during the American War of Independence, at Saratoga on 16 October, 1777, but he was a much abler soldier than that unhappy event might indicate, as well as being a man of many parts.

Burgoyne was born in London in 1722. He was the second son of a Bedfordshire baronet who lived in fine style but who died in a debtor's prison. In later years Horace Walpole spread the rumour that John Burgoyne was the natural son of Lord Bingley, at one time ambassador in Spain, who was Burgoyne's godfather, and who left Burgoyne's mother a considerable legacy. Walpole was a notorious scandalmonger and nothing has been found to substantiate his charge. Burgoyne was educated at Westminster School where he became friendly with Lord Strange, son of the Earl of Derby, which in turn led to his forming a relationship with Strange's sister, Lady Charlotte Stanley. Burgoyne could hardly have been considered the perfect match for the daughter of one of the most powerful men in England, which resulted in the infatuated couple eloping to France where they were married in 1751. At the time of his elopement Burgoyne was serving in the 1st Royal Dragoons (the Royals) but he had been compelled to escape his creditors by fleeing to France. In 1756 he returned to Britain and purchased a Captaincy in the 11th Dragoons, presumably with money provided by the Derby family. War broke out soon afterwards with France and Burgoyne distinguished himself in the operations against St Malo and Cherbourg. This presumably brought him to the notice of King George II who fancied himself as a soldier; he was the last British sovereign to take the field in person — at Dettingen in 1743 where he proved to be more of a nuisance than an inspiration.

Already well known in London society for his handsome appearance, his social graces, his daring as a gambler, and his connection with the Derby family, it was not long before Burgoyne was promoted into the 2nd Foot Guards (Coldstream) as a Captain and Lieutenant-Colonel. He was then aged 36, and when it was decided to raise two regiments of light

3   **Light Dragoon.**

This Reg.ᵗ was Created in 1759.& Commanded by Col. John Burgoyn.

cavalry, Burgoyne's was a name that came naturally to mind; he was not the kind of man who remained shyly in the background. On 4 August, 1759, he was appointed Lieutenant-Colonel Commandant of the 16th Light Dragoons and directed to raise four troops of light cavalry in the general area of Northampton. There would seem to have been no problem in obtaining either officers or recruits and there were no less than three Honourables and three Baronets on the regimental list within six months of the Regiment's formation.

Burgoyne must have drafted his recruiting poster himself, since it bears all the evidence of his pen:

'You will be mounted on the finest horses in the world, with superb clothing and the richest accoutrements; your pay and privileges are equal to two guineas a week; you are everywhere respected; your society is courted; you are admired by the Fair, which, together with the chance of getting switched to a buxom widow, or of brushing a rich heiress, renders the situation truly enviable and desirable. Young men out of employment or uncomfortable, "There is a tide in the affairs of men, which, taken at the flood, leads on to fortune"; nick in instantly and enlist.'

It is hardly surprising that the author of this extravagantly worded advertisement should in the course of time be the author of a play that took London by storm. Burgoyne was never a man to do anything by halves.

Many colonels of regiments in the eighteenth century regarded them as being their personal property and acted accordingly; although many of them were absentee landlords, appearing only at the occasional review or to display their uniform, and themselves, at court. They were content to draw the pay for the men listed on the muster rolls, many of them deserters or in some cases dead, and leave it to their subordinates to lick the soldiers into shape. Discipline was brutal; in Boston in 1770 the British soldiers of the garrison were known as the 'Bloodybacks' on account of the frequent use of the lash. A great many officers were perfectly willing to risk their lives in battle at the head of their men but considered they had better ways of passing their time than attending drill

An unknown officer of the 16th Light Dragoons. The badge on his crossbelt bears Queen Charlotte's cypher. Portrait by Thomas Gainsborough

The Lone Patrol. Oil painting by W. B. Wollen

Colonel John Burgoyne
National Portrait Gallery

parades or field manoeuvres. Burgoyne was one of the
exceptions. He had very strong views on the relationship of
officers with their men as is clear from the Code of Instructions
he issued to all his officers.

He began by comparing the systems of discipline employed
in the Prussian and French armies. In the former men were
trained like spaniels, 'by the stick'. The French substituted
honour instead of severity. 'The Germans are the best; the
French, by the avowal of their own officers, the worst
disciplined troops in Europe. I apprehend a just medium
between the two extremes to be the surest means to bring
English soldiers to perfection.' Burgoyne went on to explain
'why an Englishman will not bear beating so well as the
foreigners in question'. He emphasized the need to treat
soldiers as 'thinking beings' and insisted on his officers 'getting
insight into the character of each particular man'. Swearing at

soldiers was forbidden and some relaxation of the strict officer-soldier relationship was permissible from time to time. 'There are occasions, such as during stable or fatigue duty, when officers may slacken the reins so far as to talk with soldiers; nay, even a joke may be used, not only without harm but to good purpose, for condescensions well applied are an encouragement to the well disposed, and at the same time a tacit reproof to others.' It may seem surprising today that officers had to be encouraged to talk and joke with their men, but this was radical stuff at the time. What is equally surprising, however, is that this relaxed relationship between officers and soldiers has remained a feature of Burgoyne's regiment from that day to this.

On parade Burgoyne required proper subordination among officers, but off parade was another matter entirely. There should be complete social equality in private intercourse, and 'any restraint upon conversation (off parade), unless when an offence against religion, morals, or good breeding is in question, is grating.' He went on to require his officers to study their profession, pointing out that 'A short space of time given to reading each day, if the books are well chosen and the subject properly digested, will furnish a great deal of instruction.' He urged his officers to learn French (shades of the Common Market!), because the best military manuals of that time were written in French, and to study mathematics because it was an essential subject for the military profession. 'An officer ought to write English with swiftness and accuracy,' and 'If a man has a taste for drawing, it will add a very pleasing and useful qualification; and I would recommend him to practise taking views from an eminence, and to measure distances with his eye. This would be a talent peculiarly adapted to the light dragoon service.'

Horsemanship was more or less taken for granted among officers who had ridden since childhood, but this was not good enough for Burgoyne. Horsemanship was to be studied, as was the fitting of saddles and bridles. 'I hope I shall not appear finical, if I recommend to officers sometimes to accoutre and bridle a horse themselves until they are thoroughly acquainted with the use of each strap and buckle.' Anticipating, no doubt, complaints from officers who regarded such instruction as being beneath their dignity, Burgoyne asked them to consider 'whether a reproof from a field officer, or, what is perhaps worse, a criticism from a judicious spectator, would not give them more pain?' He even asks his officers to acquire some knowledge of farriery, and to interest themselves in the feeding of their horses.

In all this Burgoyne was years ahead of his time. Therefore it is not surprising that his newly-raised Regiment soon became efficient and fit for active service. In April, 1761, two troops under Captains Sir William Williams and Sir George Osborne took part in the expedition to capture Belle Ile off the Brittany coast. They distinguished themselves but Williams was killed. In the following year the Regiment embarked for Portugal. The Portuguese, under threat from Spain and France, invoked their old treaty with Britain and an expeditionary force was sent to their aid. It was commanded by the Count of Schaumberg-Lippe.

The 16th Light Dragoons sailed in May, 1762, and remained in Portugal for a year. They were to discover that the army of their Portuguese ally was a somewhat undependable instrument of war. At a banquet given in his honour by the Portuguese generals, the Allied Commander-in-Chief was astonished to find that the waiters were captains and lieutenants in the army he had come to command. Lippe held the Portuguese artillery in such contempt that he offered a prize for the gun team which succeeded in hitting the flag above his tent; to make the competition more exciting he directed that it should be held on the day he was giving a return banquet to the Portuguese generals. It must have seemed to Burgoyne and his Regiment that there was little prospect of such a heterogeneous mob defeating the armies of France and Spain, but in fact the Spaniards were little better.

The 16th first distinguished themselves at Valencia de Alcantara, just across the River Tagus in Spain. Burgoyne, with a brigade composed of British and Portuguese infantry and his own Regiment, surprised a superior Spanish force by a daring night march, culminating in the advance guard of the 16th Light Dragoons galloping into the main square, quickly followed by the rest of the Regiment. The Spanish general in command was taken prisoner, along with most of the Regiment of Seville, together with three stands of Colours. During the subsequent pursuit a detachment of six men of the 16th under a sergeant charged and took prisoner a troop of 26 Spanish Dragoons. Burgoyne waxed lyrical in describing this incident in a despatch to Count Lippe.

The exploits of the 16th Light Dragoons and their dashing Colonel were not sufficient to prevent a Spanish invasion of Portugal. A large force moved across the frontier but we are told 'the dauntless countenance of the British troops overawed their opponents'! Overawed or not, the Spaniards continued to advance and in consequence 'some retrograde movements were, however, necessary'. The 16th Light Dragoons covered the withdrawal and eventually the Tagus was crossed. A force under Burgoyne took up a position at Villa Velha in October and again succeeded in taking the Spaniards by surprise. It does not seem to have been a difficult thing to do.

In this instance the Spanish had occupied an old Moorish castle on the north bank of the Tagus and had fortified two hills on the plain of Villa Velha. Burgoyne formed a small force of 16th Light Dragoons, Grenadiers and Portuguese, and placed them under the command of one of his officers, Charles Lee, who had won his spurs in North America fighting against the Indians and the French. Lee was a remarkable man who later threw in his lot with the American colonists and had several exchanges with Burgoyne, as well as an unfortunate experience at the hands of the 16th Light Dragoons. Lee had begun his service in the infantry but became a dedicated cavalryman. After his service in Portugal he joined the Russian army for a time. 'I am to have command of Cossacks and Wollacks, a kind of people I have a good opinion of. I am determined not to serve in the line [infantry]; *one might as well be a churchwarden.*' Good cavalry sentiments which have been echoed down the years!

At Villa Velha Burgoyne repeated his tactics at Valencia de

Portrait of an officer of 16th Light Dragoons by Thomas Gainsborough *Tate Gallery*

Alcantara. Lee's force forded the Tagus, moved by forced march through the mountains and fell on the unsuspecting enemy around 2 am. The result could have been predicted. Most of the Spaniards were shot or bayoneted in their tents. The few who tried to make a stand were charged by the 16th Light Dragoons under Lieutenant Maitland and cut down almost to a man. Lee's force captured guns, horses and men with hardly any loss to themselves. Both he and Burgoyne figured prominently in the Commander-in-Chief's despatch. The war ended shortly afterwards and the 16th Light Dragoons and their Colonel returned to England with a considerable reputation. Lady Charlotte Burgoyne returned with them, for, like all good soldiers' wives, she had followed the drum to Portugal, despite her husband's protests.

On their return to England, the 16th Light Dragoons found themselves high in royal favour. Burgoyne was of course a very familiar attender at court, but it was also the case that the decision to form regiments of light cavalry had proved to be successful. King George III took his responsibilities as Commander-in-Chief of the Army very seriously and was never happier than when reviewing his regiments in the earlier years of his reign. On 20 May, 1766, he reviewed both the 15th and the 16th Light Dragoons on Wimbledon Common and was so pleased with what he saw that he commanded that henceforward the 15th should be styled 'The King's' and the 16th 'The Queen's' Light Dragoons. George III's consort was Queen Charlotte and the Regiment adopted her cipher as its badge. The connection with Queen Charlotte is further commemorated today in the engraved silver buttons worn on officers' blazers, the monogram being 'Q.C.L.' (Queen Charlotte's Lancers), worn originally as regimental hunt buttons.

Queen Charlotte
National Portrait Gallery

Caricature of an officer of the
16th Light Dragoons by
Robert Dighton

The subject of dress, or uniform, has always had an obsessive
interest for a great many soldiers. More time has been devoted
to determining the sit of a button or the correct colour of
facings than has ever been given to the business of war. Kings
in particular seem to be afflicted by this particular phobia and
George III was to prove no exception; only his son, George IV,
surpassed him in his interest in uniforms. Having two new
regiments to dress, the King was in his element, and by 1768
the 16th were dressed as follows. On their heads a helmet with
a horse-hair crest. The coats were scarlet with blue facings.
Waistcoats and breeches were white with black boots reaching
to the knee, and the cloaks were scarlet with white linings and

blue capes. The list of 'necessaries' for a light dragoon (Captain Hinde's *Discipline of the Light Horse*) is shown as: A helmet, a coat, a waistcoat, breeches, cloak, watering cap, four shirts, four pairs of stockings, one pair of boot stockings, black stock, one pair of leather breeches, two pairs of short gaiters, one white jacket, one stable frock, buckles for socks and garters, and a picker, turnscrew, pick wire, pan brush, worm, oil bottle, and necessary fodder bags. A new coat, waistcoat and breeches were supplied every two years, helmets every four years, and gloves every year. There were immensely detailed regulations concerning saddle cloths, horse furniture etc, some of the articles being inscribed with the cipher of Queen Charlotte. This was to have significance 200 years later.

Three Guidons were carried. These swallow-tailed flags consisted of the first, or King's, Guidon of crimson silk, and the second and third Guidons of blue silk. The two latter were embroidered with the Queen's cipher within the Garter and bore the regiment's motto, *Aut cursu, aut cominus armis*, which remains the same today. It can be translated as 'Either in speed, or in close combat,' and summed up, reasonably accurately, the role of the light dragoon. Its selection doubtless bears witness to John Burgoyne's classical education at Westminster School.

Queen Charlotte's cypher and the motto

The light dragoon was lavishly supplied with weapons. His principal weapon was the sword, about 36 inches long, either straight or slightly curved. Each man also carried two pistols with 9-inch barrels and a carbine 2 feet 5 inches long in the barrel; pistols and carbine were of the same calibre so that the same bullet fitted both. Neither was accurate for any distance but when discharged point-blank could inflict a fatal wound. It was usual to dismount before using the carbine.

Burgoyne returned from Portugal with his military reputation considerably enhanced. He was promoted to Colonel and a further mark of royal favour was bestowed on him on 18 March, 1763, when he was gazetted Colonel of the 16th Light Dragoons. This was an appointment he could expect to hold for life, and it carried with it many financial advantages. A regiment was its Colonel's property to some extent and was worth at least £3,000 a year [probably £50,000 a year at current rates]. For a free spender and gambler like Burgoyne, who had been elected MP for Midhurst in 1761, it was a life-saver. He concentrated on the House of Commons, leaving the Regiment to be run by its Lieutenant-Colonel.

Although King George III continued to review his Light Dragoon regiments at Wimbledon almost every year, sterner times were approaching. The storm that began in 1763 when the then Prime Minister, George Grenville, introduced his Stamp Act as a means of defraying the cost of the British garrison in North America, grew slowly but steadily into a tempest; it was not only to lose Britain her American colonies, but also to put an end to John Burgoyne's ambitions for military glory. On 18 and 19 April, 1775, the first shots were fired at Lexington and Concord in Massachusetts, and from that moment onwards the avalanche gathered pace. When it was decided to reinforce the existing garrison in New England, the 16th Queen's Light Dragoons were one of the regiments selected. They were commanded by Lieutenant-Colonel the Honourable William Harcourt (later Field-Marshal Earl Harcourt and Colonel of the Regiment from 1779 to 1830).

The 16th were sent to Boston where they had been preceded by Major-General John Burgoyne, one of the three general officers sent out to assist Lieutenant-General Thomas Gage, Commander-in-Chief in North America. However, the British evacuated Boston on St Patrick's Day, 1776, and the convoy of ships carrying the 16th and other reinforcements had to be switched to New York. The voyage had been a hideous one of more than three months in the teeth of Atlantic gales. The ships were crammed with troops, horses and supplies. The stink below decks was worse than any stable and the soldiers lived in almost total darkness for most of the time. Food was short and barely eatable. Drinking water soon went foul and deaths from dysentery and scurvy were frequent. In really rough seas the horses went mad with fear between-decks and often cast themselves. It then became necessary to fumble around in the darkness among flailing hooves to release the cast animal; and if it was too badly injured to get back onto its feet, to kill it then and there by cutting its throat. After a sea voyage of such a length and description, it is incredible that a regiment could be fit for anything, but within a week of landing in America the 16th Light Dragoons were in action against the Colonists at the Battle of White Plains on 28 October, 1776.

Cavalry were in short supply in North America and consequently the 16th Light Dragoons were given little rest. At this early stage in the war the Colonists had barely had time to organize an army and spent most of their time trying to avoid battle with the British. General George Washington was a master of guerrilla tactics, exasperating his opponents by managing to slip away just as the net was closing around him. The result was a series of encounter actions involving, inevitably, the cavalry. In November, 1776, the Colonists abandoned New York and fell back across New Jersey with the British in pursuit. Their objective was Philadelphia where the Colonists had set up a government.

Oddly enough, one of the most enterprising American leaders at this time was Charles Lee, the same Charles Lee who had been sent across the River Tagus by Burgoyne to attack the Spaniards at Villa Velha fourteen years previously. Lee was a 'loner' who quarrelled with most people sooner or later. Moody, slovenly and ill tempered, Lee was daring at one moment and fumbling the next. He had emigrated to North America and when war came volunteered to serve the Colonists. They made him a general and he was a good one.

Hamilton delin.

Hawkins sculp.

The American General Lee *taken Prisoner by* Lieutenant Colonel Harcourt *of the* ENGLISH ARMY, *in Morris Country, New Jersey, 1776.*

Early in December he was operating north of the Delaware River with a force of around 3,000 men and a few guns. After carrying out a reconnaissance, he put up for the night in an inn near Morristown in New Jersey. He had earlier boasted that he was 'going into the Jerseys for the salvation of America,' but a patrol of his old regiment fell in with a countryman and learnt that Lee was close by. At 10 o'clock on 13 December, 1776, having written despatches and breakfasted, Lee came down to mount his horse, only to be taken prisoner by the 16th Light Dragoons, whose Commanding Officer had served with Lee in Portugal. It must have been a strange reunion!

Harcourt wrote that he had captured the 'most active and enterprising of the enemy generals,' and it was thought at one time that Lee's capture might result in wholesale American defections; but it did not happen. Lee was in grave danger at first of being shot for treason but in the end he was exchanged for a British general taken prisoner in Rhode Island. He continued moody and cantankerous until his end in 1782, quarrelling with Washington and the Congress, and failing to make proper use of his undoubted abilities as a soldier; a strange, tormented creature.

For the next three years the 16th Light Dragoons marched and counter-marched through the Jerseys and Pennsylvania. They were present when the British entered Philadelphia, and they took part in the battles at Brandywine Creek and Germantown in 1777. Scouring the countryside for supplies, they annoyed both loyalists and rebels by grazing their horses in the standing corn. It was very hard campaigning because it was difficult to distinguish friend from foe. Moreover, the eastern states at that time were one vast forest, for the most part trackless, and where tracks did exist they were clogged with dust in summer and bogs in the fall and winter. It was a vast area in which to campaign and Howe, the British Commander-in-Chief, had far too few troops to hold down the settled parts, interspersed as they were among a primaeval wilderness.

Although the British affected to despise the Colonists, it was clear to some of them that the Americans were formidable opponents. As Harcourt wrote to his father in 1777:

'Though they seem to be ignorant of the precision, order, and even of the principles, by which large bodies are moved, yet they possess some of the requisites for making good troops, such as extreme cunning, great industry in moving ground and felling of wood, activity and a spirit of enterprise upon any advantage. Having said this much, I have no occasion to add that, though it was once the fashion of this army to treat them in the most contemptible light, they are now become a formidable enemy.'

The effect on General Howe seems to have been a kind of paralysis. He had acquired a mistress, the wife of his commissary of prisoners, and was enjoying with her the amenities of New York while the troopers of the 16th Light Dragoons were on outpost duties in the Jerseys, struggling to survive on hard tack in the wilderness. Howe's dilly-dallying is well described in some lines of doggerel which were popular at the time:

> Awake, arouse, Sir Billy,
> There's forage in the plain,
> Ah, leave your little Filly,
> And open the campaign.
> Heed not a woman's prattle
> Which tickles in the ear,
> But give the word for battle
> And grasp the warlike spear.

In the summer of 1777 the British marched an army down from Canada to join forces at Albany on the Hudson River with the troops already operating out of New York under Generals Howe and Clinton. A successful outcome to this campaign would cut off the New England colonists from their confederates in the Jerseys and Pennsylvania. The command was given to Major-General John Burgoyne. He had a well-found force of British and German troops but war in the North American wilderness was vastly different from that practised on the plains of Germany and the Low Countries. Logistics as much as anything defeated Burgoyne. He was a capable general, possibly one of the most imaginative on the British side, but he was eventually halted at Saratoga by superior American forces on 17 October, 1777. If not *the* turning point in the war, at least it was one of the most significant markers of the outcome of the struggle.

France entered the war on the side of the Colonists soon afterwards, and soon Spain did likewise. Britain did not possess the resources to deal with such a powerful combination of enemies. Before the end came with Cornwallis's surrender at Yorktown on 19 October, 1781, four years almost to the day since Burgoyne had surrendered at Saratoga, the 16th Light Dragoons had been sent home in 1779. They had lost nearly all their horses, which were virtually irreplaceable, and their losses in men, mostly from sickness, had reduced the Regiment to a skeleton. A dismounted detachment was sent to the West Indies as part of the force which captured St Lucia in 1778. The rest of the 16th Light Dragoons returned home from as arduous a campaign as the Regiment has ever taken part in.

On their return home they learnt that General Burgoyne had resigned his Colonelcy of the Regiment in protest against his treatment by the British government after his surrender at Saratoga. 'Burgoyne's Light Horse' would have to find another nickname. The Honourable William Harcourt was given the Colonelcy by George III and held it for no less than 51 years, by which time he had risen to the rank of Field-Marshal. Earl Harcourt, as he subsequently became, died on 18 June 1830, aged 87. He had served 71 years in the Army, no less than 62 of them with the 16th Light Dragoons. 'Gentleman Johnny' Burgoyne rose to the rank of Lieutenant-General after his return from America. He became an M.P. and was a prolix if pompous orator in the House of Commons, taking a prominent part in the impeachment of Warren Hastings. After the death of his wife Charlotte he acquired a mistress, Susan Caulfield, an actress or singer, who bore him two sons and two daughters. One of the sons, named after Charles James Fox, later became Lord Raglan's Chief Engineer in the Siege of Sebastopol in the Crimean War. He rose to the rank of Field-Marshal and was created a Baronet. His statue in Waterloo Place in London is sometimes mistaken for his father's.

The Right Honourable The Earl Harcourt

CHAPTER THREE

Peninsula

IN 1789 the fire which had long been smouldering in France broke into flames. Soon the rest of Europe became uneasily aware that the most civilized and powerful of all the European nations was in the grip of revolution. The British had little cause to love the French, with whom they had been at war on and off for hundreds of years, but revolution was another matter. If Louis XVI was to lose his head today who could say when it might be George III's turn? First Austria and Prussia, and then more tardily Britain, joined together to support the royalist cause, and in the Spring of 1793 an expeditionary force of 10,000 men embarked from England for the Continent. It marked the beginning of a war which was not to end until Napoleon was defeated at Waterloo nearly a quarter of a century later.

The 16th Light Dragoons, their strength augmented by raw recruits who scarcely knew how to ride, formed part of the force under King George III's second son, the Duke of York, of whom the well-known jingle was written:

> The Noble Duke of York,
> He had ten thousand men.
> He marched them up to the top of the hill,
> And he marched them down again!

The 16th Light Dragoons were commanded by Lieutenant-Colonel Sir Robert Laurie, a baronet who had paid around £5,000 at the regulation price for the privilege of commanding the Regiment. They marched from Ostend to Tournai where they joined forces with an Austrian army already assembled there. There followed a long and inconclusive campaign against the armies of Revolutionary France. The French levies may have been undisciplined but they made up for this by their revolutionary fervour. They even fought at night, unheard of in those days, and they did not stop fighting when winter arrived and the other European armies went into winter quarters until the spring. It was during the fighting along the Franco-Belgian frontier that the 16th Light Dragoons won their first Battle Honour, at Beaumont near Le Cateau on 26 April, 1794. Although the Duke of York commended the conduct of the 16th Light Dragoons and a regiment of German Hussars as being 'beyond all praise', it was not an action of much importance. However, it proved to be the first of the Regiment's many Battle Honours.

The skirmishes along the frontiers of France were inconclusive affairs but the Duke of York was finally compelled to retreat to the safety of the walls of Tournai, where he was attacked by the French on 10 May, 1794. More by luck than good management, of which there was an almost total absence throughout the campaign, the French were beaten off. The 16th Light Dragoons distinguished themselves by charging an artillery battery and carrying off the guns. This spirited little action gained them another Battle Honour – Willems. From then on the campaign went from bad to worse, notable as much as anything for the fact that it was during this campaign that the young Arthur Wellesley, later the Duke of Wellington, won his spurs in command of the 33rd Foot, now the Duke of Wellington's Regiment. When asked many years later if he had found the campaign useful, he replied: 'Why, I learnt what one ought not to do, and that is always something.'

The unconventional tactics and revolutionary enthusiasm of the French got the Allies on the run, compelling the British to retreat to Bremen in the depths of an unusually severe winter. It led to many deaths and not a few desertions. The Duke of York went home early in the proceedings, as did several of the other senior officers. Wellington was later to say that he was hardly ever visited by the staff, let alone by the Commander-in-Chief. It was not a glorious episode in the history of British arms. The 16th Light Dragoons embarked for England in February, 1796.

From 1795-1805 the Regiment was commanded by Lieutenant-Colonel Sir James Affleck. Affleck, who was a baronet, was born in 1759. He had served in no less than five regiments of cavalry and infantry before being appointed to command the 16th Light Dragoons from the 2nd Dragoon Guards in 1795. It was, of course, common in the days when commissions in the army and steps in rank had to be purchased for most officers to have served in several regiments. Wellington had served in no less than six before purchasing the command of the 33rd. Affleck was promoted to Major-General in 1805 and was succeeded by Sir Stapleton Cotton, also a baronet, who later became a Field-Marshal. Affleck had risen to General by the time he died in 1883.

Stapleton Cotton went on to become the Duke of Wellington's senior cavalry commander in the Peninsular War and was later raised to the peerage as Viscount Combermere. He became Commander-in-Chief of the Bengal Native Army and as such commanded at the storming of the supposedly impregnable fortress of Bhurtpore (now Bharatpur) in 1826, where his old regiment, by then Lancers, distinguished themselves. They were the first regiment of British cavalry to use the lance in battle. It was about this time that an expedition to take Rangoon was being considered. The Duke of Wellington was asked who should command the expedition. 'Why, Combermere,' the Duke replied. 'But Your Grace,' expostulated his questioners, 'we always thought you considered Lord Combermere to be a fool?' 'So he is,' said the Duke, 'and a damned fool, but he can take Rangoon!'

While the 16th Light Dragoons were engaged in putting down rebellion in Ireland, hunting smugglers in Kent, and taking part in royal reviews in Hyde Park, Napoleon was steadily extending his rule over Continental Europe. Soon he had Portugal in his sights and the Portuguese asked Britain to come to their aid. After the usual series of disasters that mark the first stages of any campaign in which the British Army has been involved (which is why such retreats as Corunna, Mons and Dunkirk have subsequently become transformed into victories), the British gradually began to catch up with themselves, sort out the logistics, sack most of the generals, and promote a few new ones in their place. They built up an army based on Lisbon, and appointed the 40-year-old Sir Arthur Wellesley, new home from India, to command it. The 16th Light Dragoons joined this army on 15 April, 1809. They did not return home for five years, in the course of which they were present at seven pitched battles, suffering 309 casualties, and losing nearly 1500 horses.

Britain never fielded an army with the exuberance, confidence and sheer professionalism of Wellington's Peninsular Army until it came to the Western Desert (in Egypt and Libya) nearly 130 years later. That army, the 8th, had the same *élan* and *panache* as Wellington's, with better discipline when let loose in Tripoli perhaps, but with the same 'base-wallahs' haunting the cafés and hotels of Lisbon and Cairo. And both armies were led by men with a true genius for war, who could not have been more alike in some ways, nor less alike in others. 'Old Nosey' the soldiers called one of them; the other they called 'Monty'. They were as much terms of endearment as anything else.

The 16th Light Dragoons began the campaign under the command of Lieutenant-Colonel the Honourable George Anson, a son of Viscount Anson (later the Earl of Lichfield). One of his subalterns was William Tomkinson, son of a Cheshire squire, whose diary of those eventful years is one of the best accounts of life in a cavalry regiment during the Peninsular War. Tomkinson was described by a brother officer as 'one of the best soldiers I have ever met'; and although he was a countryman at heart, always longing to get back to his estate at Tarporley in Cheshire, he was also a keen student of his profession. In this he was unusual since most cavalry officers of his generation believed that their principal task in war was to charge the enemy and die sword in hand.

The Battle of Talavera

The newly constructed Madrid to Lisbon motorway passes through the Talavera battlefield. In 1990 the 8th Duke of Wellington inaugurated a memorial on the site. 16th Light Dragoons is inscribed on one of the arches

Wellington's campaign in Portugal and Spain lasted for five years, with fluctuating fortunes. By the end of it all the regiments involved had reached a peak of efficiency unequalled in the British Army since Marlborough's days. The 16th Light Dragoons' first action was at Albergueria Nova on 2 May, 1809, and from then onwards they were employed almost continuously, principally on outpost duties and escorting convoys. They took part in the battles of Talavera, Fuentes d'Onor, Salamanca, Vittoria and Nive, all of which are emblazoned on their Guidon, as is Peninsula. They covered Wellington's withdrawal to the Lines of Torres Vedras in the autumn of 1810, and followed up the retreating French army in the Spring of the following year. 'We soon learnt to sleep in the day or night,' wrote Tomkinson. 'We never undressed and at night all the horses were bridled up, the men sleeping at their heads, and the officers of each troop close to their horses.' At Bienvenida (after Wellington's capture of Badajoz in April, 1812), there was a spirited clash between the cavalry of both sides. The 5th Dragoon Guards, greatly out-numbered, made two charges, taking the enemy in the flank, while the 16th Light Dragoons, jumping a wall in line, galloped head-on into the enemy. It might have proved to be a famous victory but Cotton, who was commanding the cavalry, failed to pursue the shattered enemy.

It was a high-spirited army, in which the cavalry and the Rifles set the pace for gaiety. When they were not fighting there were fox hunts and balls, as well as much wining and dining and courting of the senoritas. For weeks on end the cavalry slept out in all weathers on the outpost line, lucky if they could find temporary shelter from the snow or rain in a shepherd's flea-ridden bothy. The officers had to be tough because the men they commanded were astonishingly so. One soldier had a leg and an arm amputated without any kind of anaesthetic, and was found the following morning propped up on his remaining elbow calmly smoking his pipe. The women who followed their husbands on the line of march were equally tough. One of them, Bridget Skiddy, carried her man, knapsack, musket and all, when he could march no farther during a retreat.

Captain Tomkinson must have been equally tough. At Grijo on 11 May, 1809, the 16th Light Dragoons were advancing down a deep and narrow lane under fire from some 3,000 French infantry. They charged the enemy who ran. Tomkinson was shot in the left arm and his horse had a bayonet stuck into it. This made it virtually unrideable but Tomkinson kept his seat, although by then wounded in four places, 'a musket shot through the neck, another through the right arm above the elbow, and a third through the left fore-arm, with a bayonet thrust close to the last.' His horse eventually unseated him by knocking Tomkinson's head into a low branch. Despite his wounds and concussion, Tomkinson eventually recovered, as did his horse, which he rode at Waterloo.

Possibly the most universally admired of the Regiment's officers was the Honourable Edward Somers-Cocks who commanded a troop in the early stages of the campaign. He was killed in the breach at the storming of Burgos in 1812, by which time he had transferred to the 79th Highlanders (later the Queen's Own Cameron Highlanders which amalgamated in 1961 with the Seaforth Highlanders to form the Queen's Own Highlanders). He was the son of Baron Somers, and had the distinction of actually being mentioned in one of Wellington's Despatches, a rare honour because Wellington was very sparing in his praise. Tomkinson wrote of Somers-Cocks, 'The men in his troop were very fond of him, and would hollo when in a charge, "Follow the Captain! Stick close to the Captain!" ' They called his squadron the 'Fighting Squadron'. His funeral was attended by Wellington and nearly all the officers in the army who were not on duty. Wellington said of him; 'If Cocks had outlived this campaign, which from the way he exposed himself was morally impossible, he would have been one of the first generals of England.'

Another regimental character was Sergeant-Major Blood who behaved so gallantly at Tudela that Wellington offered him an immediate commission. Blood refused it and Wellington gave him 100 dollars instead. After the French defeat at Vittoria, Blood and six of his men managed to loot 6,000 silver dollars. He later obtained a commission as a Riding Master. There is a

The Battle of Grijo

memorial to him in Cheadle Church which was subscribed to, among others, by Lord Combermere and the Duke of York, the last saying, 'H.R.H. thinks it right to state that Mr Blood is one of the most meritorious old officers in the King's service.'

Cavalry officers used to assert that the main purpose of cavalry was to give tone to what otherwise would be simply a vulgar brawl. This did not endear them to the rest of the army. 'Perhaps I need not tell the reader,' wrote Gleig, 'that between the infantry and the cavalry of the British Army a considerable degree of jealousy exists, the former description of force regarding the latter as little better than useless, the latter regarding the former as extremely vulgar and ungenteel.' Wellington used to complain that his cavalry could 'gallop, but could not preserve their order'. However, he had no cause to complain of their handling of outpost duties, as is plain from the

The Scarlet Lancers

*History of the Sixteenth, the Queen's Light Dragoons (Lancers)
1759 to 1912* by Colonel Henry Graham, and published in
1912. Writing of the Regiment's record in the Peninsula,
Graham states:

'The history of the Sixteenth Light Dragoons in the Peninsular
War shows most clearly both what Light Cavalry can do, and
what it ought to do. The Regiment never lost an opportunity
of charging the enemy, whether infantry, cavalry, or guns, and
on every occasion it acquitted itself with honour. It was, it is
true, sometimes obliged to retire before superior numbers, but
its magnificent discipline on every occasion brought it off
without disorder and without disgrace. Nor was the Sixteenth
less successful in the other and no less important duties of Light
Cavalry. When the Regiment formed the picquets the army
behind it slept secure, for in no single instance were its outposts
surprised; when it was on reconnaissance no General need
hesitate for want of information; when employed to harass the
enemy's outposts no French picquet rested in peace. In advance
or in retreat, in quarters or in the field, the conduct and
discipline of the Sixteenth during the Peninsular War was all
that could be desired and the Regiment gained honour and
distinction for every officer who was fortunate enough to
command it.'

A proud record indeed!

The Battle of Salamanca 22nd July 1813

CHAPTER FOUR

Waterloo

THE 16th Light Dragoons returned from France in July, 1814, and were stationed at Hounslow. Soon afterwards they found themselves employed in dealing with the London mobs out in the streets and rioting in protest against the passage through Parliament of the Corn Laws. Heroes at one moment as they returned from the wars, they were now heartily booed as they rode through the streets on their way to Westminster. Aid to the Civil Power has long been regarded in the British Army as one of the most distasteful of its many duties, and it must have come as some relief when the news arrived that Napoleon had escaped from Elba on 1 March, 1815, and that the war with France was about to begin all over again.

As has happened so often after all her wars, Britain's first action had been to start dismantling her army. Consequently it took longer than it need have done to assemble an army and transport it to the Continent; but on 11 April, 1815, the 16th Light Dragoons embarked at Dover and Ramsgate for Ostend. The transports were small colliers ill-suited for the carriage of horses but fortunately the sea was calm. On arrival at Ostend the animals were slung overboard and left to their own devices to swim ashore. They were escorted by troopers who landed naked in full view of the scandalized Belgian families who were enjoying themselves on the beach.

On 2 March, 1815, the Regiment had been joined by Lieutenant John Luard who had served throughout the

Peninsula campaign with the 4th Dragoons. The reduction in the army which had followed had placed Luard on half-pay, from which he was rescued by his father's purchasing for him a Lieutenantcy in the 16th Light Dragoons at a cost of £2,000. Luard was to serve with the Regiment for 17 years.* He was an artist of some distinction who began his service in the Royal Navy, narrowly missing being present both at the battles of Trafalgar and Waterloo; his ship was on the West Indies station in October, 1805. The Spaniard, General Miguel de Alava, who served on Wellington's staff in Spain and at Waterloo, is probably the only man to have taken part in both battles.

The 16th Light Dragoons soon discovered that campaigning in the Low Countries was a distinct improvement on Spain and Portugal. 'We spent our time visiting the old churches and museums,' John Luard recorded in his diary, 'and there were many balls given in our honour.' The soldiers were similarly fêted, according to Tomkinson. 'The men cannot stand the good treatment they receive from the persons on whom they are billeted,' he wrote, 'and some instances of drunkenness have occurred.' Lieutenant-Colonel James Hay, who commanded the Regiment, took his officers to visit the battlefield of Oudenarde. He asked Luard for his comments on the battle, which was not one Luard knew anything about. 'My answers were at best guesses,' he said. 'Since the colonel did not question them I can only conclude he knew little more about the battle than I did.'

During April and May the allied armies concentrated in Belgium under the Duke of Wellington. Little news came out of France. 'We heard little information,' wrote Tomkinson, 'and the present distribution of the troops appears more like a distribution for winter quarters than for an approaching campaign.' Major-General Sir John Vandeleur, who commanded the cavalry brigade consisting of the 11th, 12th and 16th Light Dragoons, had arranged a field day for 16 June. The 16th's adjutant was sick and John Luard had been appointed to act in his place.

He was wakened at 5 am on 16 June. The Colonel wanted to see him immediately. Mounting his horse, he set off at a sharp canter to Denderwinche where regimental headquarters had

---

* John Luard's Life has been told in *Scarlet Lancer* by James Lunt, published in 1964 by Rupert Hart-Davis (in New York by Harcourt, Brace & World).

Stylised self portrait
by Private Charles
Stacey sent to his
family, proudly
displaying the
Waterloo battle
honour and his
medal

been established in an inn. On the way he passed Tomkinson
riding hard in the direction from whence Luard had come. He
told Luard that he had fallen in with a Belgian dragoon in
Enghien carrying despatches from the frontier. The French
were on the move towards Charleroi and the Prussians were
advancing to meet them. Hay had already heard the news and
on Luard's arrival ordered him to assemble the Regiment. It
was midday, however, before the 16th Light Dragoons received
their orders. They were to march to the crossroads at
Quatre-Bras. By 2 pm they had reached Nivelles where they
met the first wounded and stragglers from the battle. Now they
could hear the booming of the guns and the sound caused Hay
to order the Regiment to trot. Then Hay, taking Luard with him
and following after Vandeleur and his staff, went cantering
ahead towards Quatre-Bras. But the light was fading by the time
the 16th reached the battlefield; there was little to see except
the occasional flash of musket or cannon, and the winking fires
of the outpost line. The night air stank of gunpowder and the
ground was littered with the breastplates of the French
Cuirassiers. Time and again they had charged gallantly, and
been thrown back equally gallantly, by the rocklike squares of

the 42nd, 79th and 92nd Highlanders.

John Luard slept that night in a cabbage patch a few miles north of the crossroads. Around him lay the weary troopers wrapped in their cloaks, stiff and sore after twelve hours in the saddle. Their horses were equally weary, hardly able to summon up the strength to nibble at their hay. Both horses and men were thirsty. There were no streams and the few wells had long since been drained dry. The next morning dawned dull and lowering. Luard rode down the road towards the outposts. There was no sign of movement from the French and the smoke of their bivouac fires showed they were busy preparing *petit déjeuner.* The enemy had still made no move by ten o'clock and Wellington himself rode round the outposts to see things for himself. But meanwhile the main body of the British Army, protected by a screen of cavalry outposts, was withdrawing to a previously selected position at Waterloo, twelve miles north of Quatre-Bras and covering Brussels.

It was not until the early afternoon that Napoleon woke up to what was happening. He was furious with Ney for failing to press hard on the heels of the retreating British. After first sending d'Erlon with his cavalry corps to break up the British withdrawal, he eventually galloped forward himself. But he was too late. At 3 pm it began to rain in torrents, turning the ground into a bog and quenching the fires of the French pursuit. 'It became impossible for the French cavalry to press our columns in any force,' wrote one of the British rearguard. 'In fact, out of the road in the track of our own cavalry the ground was poached into a complete puddle.' Gradually the pursuit tailed off, and as night fell the drenched and saddle-sore cavalry passed through the equally wet infantry, drawn up on the high ground to the south of the village of Waterloo.

His duties as adjutant kept Luard busy until long after midnight. While on his rounds he came across a soldier leaving a cottage with a grandfather clock on his back. He asked what the man intended to do with it. 'If you come to our troop,' the soldier replied, 'you'll soon see what I will do with it. I'll make the beggar tick.' As Luard passed that way later he was invited to warm himself before a blazing fire of which the clock formed the principal ingredient. When at last Luard was able to lie down in his cloak, he found it difficult to sleep: 'My breeches were soaked and the discomfort kept me awake for most of the night, although I was extremely fatigued.'

It continued to rain fitfully throughout the night. Then, as the grey sky slowly cleared, a watery sun came out, and Sunday, 18 June, began. Unshaven, bedraggled, covered in mud, stiff in every limb, the 16th Light Dragoons struggled painfully to their feet and went off to feed their horses. It had been a night to remember. 'At nine o'clock I went to Colonel Hay,' wrote Luard, 'and he ordered me to mount the Regiment, which I did. My horse was so cold and shivered so much that he could hardly stand for me to mount. The Regiment was placed on the left of the position, and we dismounted under a rising ground, waiting for orders.'

The enemy could be seen quite clearly. Teams of horses were struggling forward to position the guns. The sight proved to be too much for a veteran named Price, the shoemaker in Tomkinson's troop. Dismounting rapidly, he hurried off to the rear. No one tried to stop him and he did not rejoin until after

Corporation of London

Dinner
in Honour of the Regiments
which fought at
The Battle of Waterloo

GUILDHALL: 21 JUNE 1965

Representatives of the Regiment were invited to a dinner in the Guildhall to mark the 150th anniversary of the battle

the battle. He was a regimental character and his comrades bore him no malice. At eleven o'clock the Regiment was moved to the east of the Brussels highway. To their front the rising ground was occupied by infantry and very little happened for the next hour or so. The French cannonade began at 11.35 am but the ground was so sodden that many of the shells were buried as they landed.

The Battle of Waterloo was fought within a remarkably small area. From north to south it extended for less than four miles, and was rather less in width. In this restricted space there were deployed more than four hundred pieces of artillery and over 140,000 men. Napoleon launched attack after attack against the British lines drawn up along the ridge. Each assault was repulsed. The 16th had mounted and moved off to break up the retreat of Durotte's division after one of these attacks. They were recalled in order to charge with the rest of Vandeleur's brigade in an attempt to rescue Ponsonby's brigade of heavy cavalry. Ponsonby's — the Royals, Greys and Inniskillings — had

The closing stages of the Battle of Waterloo

galloped away out of control after charging and breaking up the retreating French. As they attempted to rally, they were attacked by French Lancers intent on their annihilation. Colonel Hay's trumpeter sounded the Charge and the 16th Light Dragoons plunged forward down the slope into the smoke-filled valley below. Luard's horse stumbled, but he collected it by main force and galloped on. At the same moment he saw his Colonel pitch forward out of the saddle. Swerving to avoid him, Luard pressed on to cut down a French lancer looming out of the smoke. Swinging his horse to avoid a bayonet thrust, Luard found himself part of the general *mêlée* as the battlefield dissolved into a series of individual combats as English light dragoon fought French lancer, each isolated from his comrades by the smoke, the noise and the general confusion. The French broke away and the 16th then rallied to the calls of their trumpeters as Vandeleur's brigade withdrew to their former position.

They brought back with them their wounded Colonel. He had been shot from behind, probably by a stray shot from his own infantry, and the wound seemed mortal. But Hay recovered and lived to become a general. The Regiment had suffered few other casualties. The smoke from the guns was now so dense that men only knew what was happening in their own immediate vicinity. At about 4 pm they were told that the Prussian army was beginning to arrive on the battlefield. At about the same time a bullet hit Luard's charger in the head and it collapsed under him. He transferred to a trooper's horse as the 16th moved west of the Brussels highway, passing the 18th Hussars in the smoke. 'I saw my brother George,' wrote Luard, 'the 18th Hussars being close to us. While in this position I was talking to Lieutenant Phelips of the 11th Light Dragoons when his head was shot off by a cannon shot.... The Belgians began to give way, the enemy's fire being too hot for them, and we closed our squadrons and would not let them go to the rear. Sir John Vandeleur and I moved to the front and encouraged them. The fire then slackened and they held their ground.' It had been a critical moment, and as so often happened, the Duke himself turned up out of the smoke to shore up his battered line. 'That's right, that's right!' he called out to Luard, 'Keep them up! Keep them up!' And placing two fingers to his cocked hat, the Duke of Wellington, 'his face blackened with smoke, but otherwise his appearance as neat as ever,' galloped off to put heart into his soldiers by his very presence.

Napoleon was about to try a last gambler's throw. The flower of the French Army, the Imperial Guard, advanced to win the day for their Emperor. At their head was a hatless *Maréchal de France*, Ney, as smoke-blackened as Wellington; and above them waved the Colours emblazoned with victory after victory, Marengo, Austerlitz, Jena, and all the other victories. They were met by controlled volleys from the First Guards and 52nd Light Infantry that broke their ranks. Their steady, measured advance faltered, hesitated, then finally recoiled. *La Garde recule!* The French infantry, watching astonished as the Imperial Guard fell back, joined in the rearward movement themselves, and as they retreated across fields littered with the débris of battle, Wellington trotted forward to give the signal for a general advance.

The 16th Light Dragoons moved to the front. The trumpets sounded. Vandeleur's Brigade poured down from the ridge. At first there was no one to sabre; the French were falling back too rapidly. But soon they came up against formed bodies of troops, some of whom had plenty of fight left in them. 'The enemy's infantry behind the hedge gave us a volley,' wrote Tomkinson, 'and being close to them, and the hedge no more than some scattered bushes without a ditch, we made a rush and went into their columns... they running away to the square for shelter.' Tomkinson led his troop in several such skirmishes as the French withdrew. In one of them Captain Buchanan fell, in another Lieutenant Hay, whose body was never recovered. He may have fallen in a cornfield and been left undiscovered until after dark, when the local inhabitants, aided and abetted by the camp-followers, crept round the battlefield stripping the dead and dying, and removing anything of value they may have had on them.

As night fell most of the Regiment flopped down beside their horses; both men and animals were too tired to eat, and there was no water. The British left it to the Prussians to continue the pursuit. The next morning a passing 18th Hussar told John Luard that his brother George was safe. 'It is a curious fact,' says Luard, 'that throughout the Peninsular War, altho' we were between us in every general action from Talavera to Toulouse, and in several cavalry affairs, we were neither of us wounded; at Waterloo we both had our horses shot.'

On 19 June Wellington wrote an account of his victory to the Honourable William Wellesley-Pole. 'My dear William,' he said, 'You'll see an account of our Desperate Battle and victory over Boney! It was the most desperate business I ever was in; I never took so much trouble about any battle; and never was so near being beat.' It had indeed been a desperate business. To the end of his long life John Luard always claimed that his proudest moment was when he charged with the 16th Light Dragoons at Waterloo. It was a famous victory.

CHAPTER FIVE

From Light Dragoons to Lancers

THE 16th Light Dragoons were not chosen for occupation duties in France, returning to Romford in Essex. Cavalry regiments at that time were constantly on the move, and frequently split up into small detachments. There was a good deal of industrial and agricultural unrest, and in the absence of a properly constituted police force, the soldiers were constantly being called upon for aid to the civil power. Furthermore there was the usual reduction in strength that invariably followed the ending of hostilities. No fewer than eight cavalry regiments had been disbanded by 1822. None of this made for efficiency, nor encouraged men to join the cavalry. As a consequence those already serving, both officers and soldiers, remained in the service for far too long, some of them being barely capable of fulfilling their duties. Bumbling old generals were rooted in the past, fearful only lest some reformer should succeed in altering regulations that were hallowed by time. When the Crimean War began in 1856 there were thirteen generals on the active list who had more than seventy years' service, and thirty-seven who had between sixty and seventy years'.

A Light Dragoon regiment's establishment varied, depending on whether it was serving at home or overseas, but in grand total it amounted to 363 all ranks; this included 28 officers and 335 NCOs and privates. (The term trooper for a cavalry soldier did not become general until the end of the nineteenth century;

troop horses were also known as troopers.) Each regiment had a Colonel and Lieutenant-Colonel, the former usually serving on the Staff, the latter in command. There were six troops, normally commanded by captains, but two or more might be linked together to form a squadron. The troop, however, was the basic organization. Squadrons were not introduced until almost the end of the century.

The 16th Light Dragoons moved to Ireland in February, 1816, and remained there for three years. It was split up into small detachments all over the south where, according to John Luard, 'Several of the officers fell in love.... Colonel Pelly married Miss French; Major Persse married a Miss Moore.... It was a very hospitable quarter and the officers of the 16th were paid a great attention.' Something of the same kind happened at Catterick between 1957 and 1959, when the Regiment returned Home after seven years overseas; many captains and subalterns found their way to the altar.

February is a bad month for crossing the Irish Sea, particularly in the days of sail, and Luard has provided us with a graphic description of the kind of problems faced by a cavalry subaltern when at sea in a gale:

'At ten o'clock the next morning a brig on board of which was Lieutenant Beauchamp and 18 horses came in — this vessel sailed before I did, and in the gale which drove us into Milford Haven the bales which divided the horses gave way and they were all thrown together. The captain of the brig said he could not be answerable for her safety unless the horses were destroyed. This was not easy to be done, they could not be shot, all the dragoons except Beauchamp and the Trumpeter were seasick, the sailors were afraid to go among the horses; so Beauchamp and the Trumpeter went below and cut all the horses' throats and they were thrown overboard.'

This is a matter-of-fact account of what must have been a terrifying business. The brig was pitching wildly in the gale, the horses terrified in the darkness of the hold as they struggled to keep their feet, while Beauchamp and his assistant were groping in the dark and up to their waists in water. A false step meant certain death beneath the flailing hooves, and there were eighteen fear-maddened animals to be dispatched. There is an equally sad *dénouement* to this story, for Beauchamp, 'a most popular officer,' cut his own throat not long afterwards. Not apparently from remorse, but shortly after marrying a Miss Ball − from which Luard leaves us to draw our own conclusions.

It was soon after their arrival in Ireland that the 16th Light Dragoons were equipped with the lance, losing their carbines, and being designated Lancers. The 9th, 12th and 23rd Light Dragoons were similarly transformed into Lancers. This change was undoubtedly motivated by Napoleon's introduction of Lancer regiments into the French cavalry, the first regiment of which was Polish, 'The Lancers of the Vistula'. The introduction of the lance was not universally welcomed in the British cavalry, many criticizing it as heavy and unwieldy as compared with the sword. It had, however, several defenders, among them Lieutenant-General Dunham Massy, who was Colonel of the 5th Lancers from 1896-1906. Writing to *The Times* on 9 May 1903, General Massy claimed that 'Both in moral and physical effect it [the lance] is incomparably superior to the sword in all situations, except close *mêlée*.' Certainly the moral effect of a line of charging lancers must have been considerable, but as the 21st Lancers found to their cost at Omdurman in 1898, it was more a hindrance than a help when engaged in close-quarter fighting with stout-hearted warriors equipped with razor-sharp swords and stabbing spears. The lance-versus-sword debate remained a fertile subject for argument and discussion in British and Indian cavalry officers' messes for many years to come.

The first lances issued were an unwieldy 15 or 16 feet but soon the length was reduced to around 12 feet. To it was attached a pennon, usually in red and white.*
The butt was carried in a leather socket attached to the stirrup, and the stave, in later models, had a loop for the arm. The introduction of this new kind of cavalry offered a wonderful opportunity to the Prince Regent (later George IV) whose obsession with military uniform knew no bounds. In the cavalry the uniforms became more and more gorgeous and less and less practicable − 'Frenchified' as the Duke of Wellington contemptuously described it. Lancer regiments copied the

---

* The last occasion when lance pennons were attached in battle was at Hashin in the Eastern Sudan on 20 March, 1885, when two squadrons each of 5th Lancers and 9th Bengal Cavalry charged the Dervishes.

dress of the Polish lancers, including the head-dress (*chapka*, Polish for a hat), while the short, double-breasted tunic was called a *ulunka*; in the German and Austrian armies the lancer regiments were called *uhlans*. As time went on the *chapka* with its drooping red and white cock-tail became more and more ridiculous; it may suit the Polish physiognomy but somehow looks ludicrous on the British. At first the uniform was blue, but in 1830 it was changed to scarlet; William IV had a penchant for this colour and at one time threatened to put the Royal Navy into scarlet! In 1846 all Lancer regiments were ordered to wear blue, the exception being the 16th whose Colonel, General Sir John Vandeleur, petitioned the Queen for his Regiment to retain scarlet. This led eventually to the Regiment's nickname by the rest of the Army, 'The Scarlet Lancers'. Somewhere around this time, too, the Regiment's title was changed, firstly to the 16th, The Queen's, Light Dragoons (Lancers), and later to the 16th The Queen's Lancers.

The Regiment returned from Ireland in 1819, first to Bristol, and later to Sheffield in 1821. There they were joined by William Elphinstone who had purchased command of the Regiment. Elphinstone had begun his career in the Foot Guards but he commanded the 33rd Foot (later the Duke of Wellington's Regiment) at Waterloo. Luard speaks of him as 'a very elegant and courteous gentleman', an opinion he might have occasion to change when Elphinstone, by then a general, commanded the forces in Kabul; but that was twenty years later. It was, however, while in Sheffield, and under Elphinstone's command, that the 16th The Queen's Light Dragoons (Lancers) ran head on into conflict with their Sovereign, George IV.

King George IV was engaged in a most unsavoury tussle with his wife, Queen Caroline, who happened to be Colonel-in-Chief of the 16th The Queen's Lancers (if no better than she should be).

England was divided between those who thought the King was behaving extremely badly and those who thought he had had every provocation to do so. Certainly Queen Caroline, in Graham's history of the Regiment, looks a regular harpy! The precise cause of the rupture was the King's determined effort to prevent his Queen from being present at his Coronation. She, naturally, was equally determined to be there. The officers of the 16th, their hearts ruling their heads, or perhaps out of pure contrariness, refused to keep their feelings secret. They invited the gentry of the West Riding to dine with them, and loudly, and perhaps with unnecessary frequency (and publicity), toasted 'The Queen'. Today it would have caused scarcely a ripple, if reported at all, but in 1821 it had reached gale force when the King came to hear of it. It is doubtful too whether the Duke of Wellington, always opposed to the army meddling in politics, and in any case profoundly contemptuous of the heart ever ruling the head, would have sympathized with the officers' espousal of Queen Caroline's cause. 'Nothing to do with them, nothing to do with them!' he would have said. But the consequences were obvious. In 1822 the 16th received orders for India, from where they were not to return for 24 years. Only one officer who embarked with the regiment at Gravesend in 1822 was still serving with it on its return in 1846. He was George Macdowell who went out to India as a junior captain, and who returned in 1846 as the Commanding Officer. One can imagine King George IV saying with some satisfaction to the Duke of Wellington, 'Serve them right, Arthur! Serve them right!'

Service in India was not very popular in regiments like the 16th The Queen's Lancers. India to begin with was a very long way away, four, five or six months via the Cape of Good Hope, and three to four overland via Suez. The journey was both uncomfortable and dangerous, and also expensive. The climate was bad, and survival was uncertain. Few families were permitted to accompany, and those who remained behind were unlikely to be reunited. Officers of the King's regiments, like the 16th, spent much of their time squabbling with officers of the East India Company's armies, over whom they claimed precedence, but at lesser rates of pay. India was the land for the ambitious, the impoverished (even the private soldier in India lived better than at home), and the seeker after military glory. But it had not much to offer the ordinary regimental officer or soldier.

Queen Caroline. Oil sketch by George Hayter. *National Portrait Gallery*

'We are off to Hindoostan,' wrote John Luard. 'What shall we find when we get there?'

The Regiment's destination was Cawnpore [now Kanpur] which they reached by river-boat up the Ganges. These boats were hauled upstream by their crews, pulling and pushing the unwieldy craft over the sandbanks. Each night they tied up by the bank and the soldiers stretched their legs in the cool of the evening. It took the 16th Lancers nearly four months to reach Cawnpore from Calcutta. No concession was made to the climate so far as dress was concerned and the high-necked scarlet tunics were black with sweat. As they travelled farther and farther into the heart of India, the country grew flatter and more dusty. 'Hotter than Hades and a damned sight less interesting,' is how one troop sergeant described it.

They were not all that much better off when they reached Cawnpore. Officers could find plenty to do during off-duty hours but the soldier was not so fortunate. 'In the first place,' writes Sergeant Thomas, 'a troop is huddled together in one barrack to the strength of 87 men. These have continual intercourse with one another as they are ranged at each side of the barrack... the duty of a dragoon is very easy on account of the great number of followers. Each man has a syce to clean and saddle his horse. Each troop has two barbers, two shoe blacks, two belt cleaners, and eight dhobies or washerwomen.... Thus the men have nothing to do except field service during the cold season. The cold months glide rapidly past and the European is recruited to health and vigour, but he must prepare early in April for his term of imprisonment — the hot winds, suffocating dust and doors closed with tatties [screens of dried grass], the barrack room being watered from outside by natives to keep them cool. At this time the men know not how to pass away the time unless by drinking or gambling; thus they are led to be drunkards or gamblers before they have been many years in India.'*

Sergeant Thomas blamed much of this moral collapse on the 'regimental canteens which deprive the soldier of his comforts while realizing a fortune in two or three years for the canteen sergeant.' He mentions one such sergeant who made £1,400 in

* In 1938, serving on the plains during the Indian hot weather, I used to marvel at the boredom of the ordinary soldier's existence in India, although this was at Amritsar where we were constantly being called out in aid of the civil power.

the space of 18 months. The mortality in India from the climate, disease, drink and vice was appalling, and yet the majority of soldiers were hardly worse off than they would have been at home. Many of them volunteered to remain in India when their regiment's tour of duty overseas came to an end. Their pay was only one shilling a day but it went further than in England, and there was in addition an allowance of two drams of rum a day. When off duty soldiers were permitted a considerable degree of freedom and many of them married Indian women. Their own countrywomen were in short supply and in great demand. A married man in H.M. 13th Light Infantry had died from cholera. His wife attended his funeral, and on leaving the cemetery was proposed to by another man in the regiment, to whom she had to reply that she was *already* engaged!

It is not surprising that both officers and men alike should welcome the opportunity of active service as a welcome break in such a monotonous existence. The 16th had their opportunity in 1825. The Rajah of Bhurtpore had seized the throne from his infant-nephew, defying the East India Company which had recognized the child's right to the throne. Bhurtpore, known as the 'Bulwark of Hindoostan', was one of the strongest fortresses in India. It was the capital of the Jats, Hindu yeomen who make excellent soldiers. They had successfully repulsed Lord Lake in 1805 when he had attempted to storm the city – no less than five times. The East India Company was determined to avenge this defeat. Lord Combermere, Commander-in-Chief of the Bengal Army, took command in person, assembling a considerable train of siege artillery. The 16th received their orders to march on 5 November. It had been a very bad hot weather in Cawnpore. Five men had committed suicide, and as the Regiment marched out of cantonments to join the Bhurtpore Field Force, a soldier drew his pistol and blew his brains out.

The force reached Agra on 3 December where it was reviewed by Combermere. The 16th were delighted to see their old commander again. They turned out at midnight to give him three cheers. He may have been stupid but he had a good way with soldiers. The 16th cheered him again on 8 December when he gave orders for the advance to Bhurtpore. The balls protecting the steel tips of the lances were removed and the 16th led the advance, arriving under the forbidding walls of Bhurtpore two days later.

'On the 10th,' wrote Luard, 'Colonel Murray, who commanded our brigade with four guns of Horse Artillery, turned out at half-past 3 a.m. The infantry remained in camp. We proceeded to Sesma, then brought our left shoulders up and led straight for Bhurtpore. I was ordered to command all the skirmishers. I was ordered by Colonel Murray to cut off any enemy I could. I led the skirmishers close under the walls, while Skinner's Horse under the command of Mr Fraser, a civilian, made a sweep to the right. Some of the enemy's horse encamped under the walls retired as we advanced, but another party encamped further out were attacked by Fraser and driven towards one of the gates of the fortress, while I galloped on with my skirmishers and intercepted them as they approached the gate of the fort. We killed and wounded about 50 and took 100 horses.... The guns of the fort now opened up.... The skirmishers were then called in. Had I been supported by Infantry, I could have galloped into the fort with the retiring enemy horse.'

Bhurtpore, 1826. A drawing by Captain J. Luard (16th Lancers)

One interesting fact regarding Luard's encounter with the Jat horsemen outside Bhurtpore is that it was the first time British cavalry had used the lance in battle. As a battle it did not amount to much but Luard could claim that he had successfully blooded his lances. It was during the night march to Bhurtpore that one of the soldiers fell down a well in the dark, horse and man together, but they were hauled up without damage. Combermere now laid siege to the fortress and the 16th had little to do. The Jats kept up a steady fire from the walls, becoming noticeably more accurate after Christmas Day when a sergeant of the Bengal Artillery named Herbert deserted to the enemy. Finally Bhurtpore was stormed on the morning of 18 January, 1826, after a mine had blown a breach in the walls. Combermere had to be forcibly restrained from accompanying the stormers.

The fortress was not surrendered until four in the afternoon and only after fearful carnage. The Jats, wrote Luard, 'fought individually to the last, yielding their guns only with their lives.' He also paid tribute to the gallant conduct of the Company's sepoy regiments. The Rajah, Durjan Sal, was captured when trying to escape, and the deserter, Herbert, was hanged from the highest battlement. Lieutenant Mackinnon of the 16th witnessed the execution and wrote in his diary: 'The numerous spectators present can bear witness to the prolonged sufferings of the culprit. The rope being adjusted, one native pushed him off a low cart under the gibbet, while two others tugged at the rope to hoist him up. The convulsive writhings of the sufferer long haunted me. They lasted nearly twenty minutes.'

The Bhurtpore Field Force was broken up a few days later and the 16th Lancers marched to Meerut. They had lost no one killed at Bhurtpore and they more than replaced their killed horses by those taken from the Jats. There was also prize money. Lord Combermere's share was £60,000, and each lieutenant-colonel received £1,500. Luard received £450 as a captain, and subalterns received £250. European sergeants were given £12 and the rank and file £4 apiece. The officers subscribed £1,000 for each of the widows of the four European officers killed in the battle; they also subscribed £1,000 to be

Camp followers at Shekoabad. Drawing by Captain J. Luard (16th Lancers)

distributed among the widows and orphans of the sixty-one European soldiers killed at Bhurtpore. By the standards of the time it was not ungenerous.

The 16th were still at Meerut in 1838 when ordered to join the 'Army of the Indus' assembling at Ferozepore, in the Punjab, for the invasion of Afghanistan. The Governor-General, Lord Auckland, alarmed at the prospect of Russian penetration into Afghanistan, decided to send an army to Kabul to overthrow the Amir, Dost Mahommed, and restore to the throne Shah Shujah, supposedly pro-British, who had been ousted by Dost Mahommed and driven into exile in British territory. It turned out to be a disastrous campaign. Dost Mahommed was a strong ruler and more popular than Shah Shujah. Although the expeditionary force reached Kabul without much difficulty, Shah Shujah proved to be incapable of governing his unruly people. Eventually the British garrison was compelled to retreat and was virtually annihilated while retreating through the passes in mid-winter from Kabul to India.

A huge number of non-combatants accompanied the fighting troops. There were men to pitch the soldiers' tents, because no European in India could be expected to pitch his own; grooms to tend the soldiers' horses, and grass-cutters to cut grass for them; drivers to twist the tails of the patient bullocks as they hauled the creaking carts laden with camp furniture, boxes of mess plate, crates of crockery, boxes of wine and cigars, and every other luxury required to make campaigning tolerable; men to lead the long strings of camels carrying hay for the horses; other men to drive the officers' buggies, to cook the food, to cut hair, to draw water, to clean the camp sites, to sell the soldiers food, drink and women. The 16th had nearly 5,000 followers to administer to their needs and they were only one of many regiments. Their officers also took a pack of foxhounds with them to fill in time when they were not fighting.

There was in fact very little fighting. Most of the hardship was due to the long marches through barren and inhospitable country in extremes of heat and cold. Kandahar surrendered without a battle. Ghazni, one of the strongest fortresses in the country, was stormed on 23 July, 1839, and Dost Mahommed fled from Kabul to the mountains. At 4 pm on 7 August, Shah Shujah, escorted by a squadron of the 16th Lancers, made a state entry into Kabul, where he was ignored by his subjects. However, the 'Army of the Indus' had not done too badly. It had marched 1,500 miles through desert and mountains; it had defeated the Afghans whenever they chose to make a stand; and it had restored Shah Shujah to the throne. The time had now come to enjoy themselves and one of the first acts was to lay out a race-course and a cricket pitch. The climate was good, the countryside pleasant, and the local ladies were handsome and in some cases forthcoming. On the surface everything seemed satisfactory, but the fires were smouldering beneath.

Kabul's amenities soon palled. Soldiers who wandered far from the camp were murdered and robbed. The 16th Lancers buried their Commanding Officer in Kabul. Lieutenant-Colonel Robert Arnold had been sick since leaving Kandahar. He was carried in a litter to Kabul where he died. His effects were

Shah Shujah, escorted by a squadron of the 16th Lancers, making a state entry in Kabul

auctioned a few days later and fetched a pretty price. A square bottle of mustard went for £7. Arnold, who was very popular, had never properly recovered from a wound received at Waterloo. The 16th lost two other officers in the campaign, as well as eighty-three soldiers and 233 horses, mostly from disease.

The 'Army of the Indus' was broken up by a General Order published on 8 October, 1839. The 16th Lancers, among other regiments, were ordered to return to India, which they did via Jalalabad and the Khyber Pass. The march took nearly four months and John Luard, by then serving on the staff in Calcutta, records with satisfaction 'the safe return of the 16th Lancers to Meerut, and with them the foxhounds they had taken with them when they set out from that place'. The only serious incident occurred after the Regiment had arrived in India. When they reached the River Jhelum, about 400 yards wide, the river was in flood. Some thirty boats had been collected to transport the baggage, and the soldiers too, if required. But staff officers on

Ghazni, the day after the attack. A drawing by Lieutenant T. Wingate

the spot said the river was fordable, although the current was swift and the stakes marking the crossing were not visible. Captain Hilton crossed the river and returned to say the river was fordable. The Brigadier then ordered the Regiment to cross and they entered the water in columns of threes. The advance guard reached the opposite bank safely but the main body lost direction and were swept away by the current. Confusion was made worse by baggage camels which entered the river higher up and were swept downstream by the current. Captain Hilton was drowned, and with him ten soldiers and twelve horses. It was an unnecessary disaster and there was much indignation with the staff for allowing it to happen. It was at this crossing that the 16th Lancers lost nearly all their officers' mess plate.*

The 16th Lancers arrived in Meerut on 18 February, 1840, having marched 2,483 miles in 463 days, of which 212 days had been spent on the march, the balance having been halts in Kabul and other places. The foxhounds must have been very footsore! In Afghanistan, however, things went from bad to worse. Although Dost Mahommed had surrendered and been sent to Calcutta in exile, his supporters, headed by his eldest son, kept the country in a state of constant tumult. The British in Kabul continued to behave as if the country was at peace, nor were matters improved when a new general arrived from India in the autumn of 1840 to command the Kabul garrison. He was Major-General William Elphinstone who had commanded the Regiment nine years earlier, since when he had prematurely aged and become a martyr to gout. He had not served previously in the east, and although John Luard liked him as a gentleman, he says Elphinstone suffered 'from want of decision and never trusting to his own opinion' which made him 'unfit to command even a regiment'. Commanding the Kabul force proved to be altogether too much for him. He was incapable of dealing with the situation when the Kabul mob rose against the British towards the end of 1841. Eventually the Kabul garrison of around 4,500 soldiers and 12,000 Indian followers retreated to India under a safe conduct from the Afghan chiefs which was not honoured. The column was massacred in the snow-covered passes and ravines, only one

* This accounts for the fact that the Regiment's silver and memorabilia do not compare with those of most other cavalry regiments. They lie at the bottom of the River Jhelum!

man, Doctor Brydon, reaching Jalalabad alive. Elphinstone surrendered to the Afghans and died in captivity. It was the worst disaster suffered by the British in the East until the surrender of Singapore almost exactly one hundred years later: and it was the historian of the British Army, Sir John Fortescue, who pronounced the final judgement: 'From beginning to end,' he wrote, 'it brought nothing but disgrace.'

The Regiment was fortunate indeed to have escaped such a disaster. Two years later, in 1843, they took part in a minor campaign against the great Mahratta chieftain, Scindiah of Gwalior. It was not very arduous, nor prolonged, but they lost two men killed and seven men wounded, chiefly from artillery fire in a battle at Maharajpore, a minor affair that was subsequently awarded them as a Battle Honour. The bronze stars awarded for the campaign were made from metal of captured Mahratta cannon.

Maharajpore Star

Every regiment has its characters, good and bad, who provide salt for the dish. The 16th Lancers had a remarkable character in John Rowland Smyth, an Irishman from County Waterford who joined the Regiment in 1821, and commanded it from 1847-55; in between he served in two other regiments. He can lay claim to some fame as having fought one of the last duels that were such a feature of the period. Smyth was serving in Dublin as a captain in the 32nd Regiment in 1830. He was called out by a Mr O'Grady whom Smyth had pulled from his horse and horsewhipped. Smyth thought O'Grady had struck him with his own whip while driving past Smyth in a cabriolet. O'Grady denied having intentionally struck him but he called Smyth out nevertheless. They met the next morning in Phoenix Park where O'Grady fell mortally wounded at the first shot. Smyth and his second were tried for manslaughter and sentenced to twelve months imprisonment. They were granted a year's leave of absence while they served it! Smyth was later transferred to the 6th Dragoons in 1839, returning to the 16th Lancers in 1842. We shall hear of him again at the Battle of Aliwal. His sentence of imprisonment certainly did not affect his career in the army. He reached the rank of Lieutenant-General and was made K.C.B. His sister, 'the beautiful Penelope Smyth', married the Prince of Capua, son of King Francis I of the Two Sicilies. Since his family opposed the marriage, the couple were married *four* times, twice in Italy, once at Gretna Green, and finally by banns at St George's, Hanover Square, London.

CHAPTER SIX

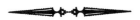

Aliwal

MOST regiments in the British Army choose to celebrate one of their Battle Honours annually as a Regimental Day. The 16th Lancers chose Aliwal, the name of an obscure Indian village, a Battle Honour it shares with only three other regiments, one of them the 2nd King Edward VII's Own Gurkha Rifles. Aliwal is in the Punjab on the south bank of the River Sutlej, one of the tributaries of the River Indus. The village consists of a few hundred mud-brick houses standing at the edge of an almost featureless plain, apart from the occasional farm surrounded by mango trees. Irrigation and modern farming has transformed the Indian countryside but in 1846 the place was virtually desert.

The inhabitants of Aliwal and the surrounding villages were almost entirely Sikhs, followers of one of the many sects of the Hindu religion. Sikhism is a militant and almost protestant religion which forbids its followers to cut their hair, worship idols or smoke tobacco. The Moslem rulers of Hindustan, the Moghuls, subjected them to merciless persecution, but without effect. The Sikhs learnt to fight hard for their faith and were excellent soldiers. They are also very hard workers, strong and sturdy, and very independent-minded. Unfortunately they are also great intriguers and extremely clannish, their clans (or *misls*) frequently at war with one another. That was their situation at the beginning of the nineteenth century when they were united by Maharajah Ranjeet Singh, a man of genius.

The unity he imposed by a mixture of force and fraud lasted no longer than his death in 1839. There were a series of palace revolutions, the efficient Sikh army got out of hand, and the Punjab rapidly degenerated into anarchy. This was watched uneasily by the East India Company which had no wish to find itself at war with the Sikhs. It had burnt its fingers badly by the disastrous war in Afghanistan and was anxious to avoid any kind of repetition. However, as the years passed, the situation in the Punjab grew more and more dangerous as the mutinous Sikh soldiers in Lahore clamoured to be led across the River Sutlej in order to liberate their co-religionists living under the Company's rule. The Company set about increasing their frontier garrisons, and at the same time began to concentrate an army of 30,000 men with which to defeat the expected Sikh invasion of British territory. This began on 11 December, 1845, when the main Sikh army crossed the Sutlej.

The 16th Lancers were at Meerut when ordered to join General Gough's force in the Punjab. By Christmas Day, 1845, they had reached Ambala, but it was not until New Year's Day, 1846, that they finally joined up with the main body. By then two fierce battles had been fought, at Mudki and Ferozeshah, and although the British had remained masters of the field their casualties had been heavy. General Gough was not much of a tactician and preferred to attack frontally, like a bulldog. 'On our arrival at the camp ground the stench was horrible,' wrote a sergeant of the 16th. 'A great many were buried within a few yards of our tents. As soon as we had pitched our camp we walked out on the field of battle to view the place and for miles around we could see the dead lying in all directions. At Ferozeshah, about three miles from our tents, the dead were lying in heaps.'

The British and Sikh armies had withdrawn a few miles from each other to lick their wounds. Gough was waiting for the arrival of his siege artillery which was being dragged slowly by teams of elephants and bullocks all the 200 miles from Delhi. The news that a Sikh force had crossed the Sutlej well to his east worried Gough since this force might well intercept the artillery train. He therefore dispatched an infantry brigade with

The town and fort of Ferozeshah. From a sketch by H. Pilleau Esq., Assistant Surgeon, 16th Lancers

orders to drive the Sikhs back across the Sutlej. Its commander was Major-General Sir Harry Smith.

Smith set out for Ludhiana on 17 January, 1846, but shortly thereafter Gough was told that the Sikh strength had been increased. He therefore sent the 16th Lancers to reinforce Smith. This meant forced marches if the Regiment were to catch up, and it was unseasonably hot. There were no roads and very little water. The horses threw up a thick cloud of dust, choking their riders, and by the time the cavalry had joined up with Smith early on 20 January, both men and horses were exhausted. Far behind toiled the baggage train, with all the mess plate. Each officer had a train of ponies and camels to carry his personal kit. With the baggage there also travelled the sick and wounded. They were carried in doolies — a kind of covered palanquin offering some protection against the sun. This long, unwieldy collection straggled along for miles, guarded only by the few who could be spared from the main body.

16th (The Queen's) Light Dragoons (or Lancers) equipped for service in India

Sir Harry Smith allowed the 16th only two hours' rest before ordering the Regiment to lead the advance towards Ludhiana. They were brigaded with two Indian cavalry regiments under Brigadier-General Charles Cureton who had joined the 16th Lancers in 1819 as Lieutenant and Adjutant. He began his military career in the militia in 1806, soon ran heavily into debt, and escaped from his creditors by faking his death by drowning. He left a bundle of clothes wrapped up on the beach and disappeared. In fact he enlisted as a private soldier into the 14th Light Dragoons under the name of Roberts, fought in most of the actions in the Peninsula and was severely wounded. He was gazetted an Ensign in the 40th Foot in 1814 in his own name, exchanged into the 20th Light Dragoons, and later transferred to the 16th Lancers with whom he fought at Bhurtpore. He commanded the Cavalry Division in Gough's army against the Sikhs, but was killed in action at Ramnagar on 22 November, 1848, while trying to extricate the 14th Light

Brigadier Charles Cureton, commander of the cavalry at the Battle of Aliwal

Dragoons from some quicksands in which they had become entangled while charging the enemy.

The march continued soon after midnight on 21 January. It was then the practice in India to march in the cool hours of the night, halting soon after 10 am and resting throughout the hot hours of the day. But forced marches were another matter and Smith pressed on despite the heat. The horses plodded through the deep sand, led for much of the way by their weary riders. Any man who fell behind was immediately killed and plundered by the Sikh horsemen who hovered in the distance on the flanks of the march. The troopers of the 16th mounted many an exhausted infantryman in their saddles or allowed them to hang on to their stirrup leathers. The exhausted column arrived in Ludhiana by nightfall, still more or less intact, but the baggage train was less fortunate. It had been under attack throughout the day, many of the sick and wounded being butchered as they

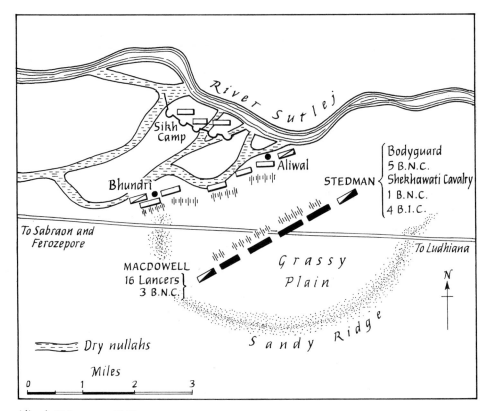

Aliwal, 28 January, 1846

lay in their litters and the greater part of the baggage looted. The 16th lost all the regimental plate they had managed to replace after the disaster at the Jhelum ford five years previously. The subaltern in charge of the baggage escort was placed under arrest by Cureton when he heard the news. He was to be court-martialled at a convenient moment, but he charged with the Regiment at Aliwal, where he was killed.*

As the British advanced, the Sikhs withdrew to the fords across the River Sutlej. Early in the morning of 28 January, 1846, the British caught up with them. Corporal Cowtan of the 16th Lancers, one of the leading scouts, has written: 'We came in sight of them about 6 am and formed into line. At this moment the view of the two armies was beautiful indeed — a fine, open, grassy plain, and the enemy in line out of their entrenchments ready to commence; the river in their rear, and in the distance the snowy range of the Himalayas with the sun rising over their tops.'

The Sikh line stretched along a ridge, about a mile long, connecting the villages of Aliwal and Bundri. About 40,000 Sikhs manned the entrenchments dug along this ridge; they were supported by 37 pieces of artillery, an arm they used with great skill and efficiency. Their flanks were protected by the Sikh cavalry. Sir Harry Smith and his staff took up position on another ridge, about 2 miles from the Sikh line. They watched from there as the infantry deployed on to the plain to their front. A cloud of dust hung over the troops as they advanced, their colours flying in the morning breeze, and with drums beating. Smith's force totalled about 10,000, of which the 16th Lancers, and 31st, 50th and 53rd Foot were British [later the East Surreys, Royal West Kents and King's Shropshire Light Infantry]. The balance was made up of the Company's native regiments, two of them being Gurkha battalions. The 16th Lancers, their lance pennons fluttering in the breeze, led the advance, with the 31st Foot close behind them, 'emulous for the front,' as Fortescue later described them. They had been charged by Harry Smith with the task of taking by the bayonet the Sikh strongpoint of Aliwal. This they did, but at heavy cost.

The loss of Aliwal jeopardized the rest of the Sikh position

* Many years later a silver-gilt cup with the crest of the 16th Lancers was found in a pawnshop in York. It has always been supposed that this cup formed part of the looted plate, but how it found its way to York must remain a mystery.

because their escape route across the Sutlej could now be cut. They immediately threw their cavalry into the battle with orders to recapture Aliwal. Suddenly a mass of horsemen emerged from behind the ridge at Bundri. Recognizing the danger, Smith at once set a galloper to order a squadron of the 16th Lancers, and another of the 3rd Bengal Light Cavalry, to charge and break up the enemy cavalry. The 3rd hesitated but Captain Bere's squadron of the 16th set off immediately without waiting for them. The light Sikh horses could not stand up to the heavy British chargers, and the British lances far out-reached the Sikh swords. They soon took to their heels. More than a thousand Sikh horsemen had been scattered by barely 100 lancers.

It was a different story, however, with their infantry and artillery. Maharajah Ranjeet Singh, convinced that the British owed their superiority on the battlefield to the way they handled their artillery, had made a point of training his own gunners along the same lines, employing European officers to teach them. The Sikh artillery was therefore always well served.

Major-General
Sir Harry Smith,
Baronet of Ailwal

The Sikh infantry, also, was extremely steady and certainly lacked nothing when it came to courage. They adopted a triangle formation; if one side of the triangle was pierced, the other two sides faced inwards and fired indiscriminately on friend and foe. Any enemy soldier who fell wounded in this *mêlée* had no hope of quarter; he was cut to pieces immediately. As Bere's squadron came galloping back from their success against the Sikh cavalry, their way was barred by one of these Sikh formations which greeted them with a volley. Bere did not hesitate but charged the enemy. He survived with a nasty wound in the face.

According to Corporal Cowtan, who charged with Bere's squadron, 'Sergeant Brown was riding next to me and cleaving everyone down with his sword when his horse was shot under him, and before he reached the ground he received no less than a dozen sabre cuts which, of course, killed him. The killed and wounded in my squadron alone was 42, and after the first charge self-preservation was the grand thing, and the love of life made us look sharp, and their great numbers required all our vigilance. Our lances seemed to paralyse them altogether, and you may be sure we did not give them time to recover themselves.'

The Sikh artillery continued to pound the British and under cover of this fire enemy infantry, supported by a battery of guns, deployed from the ridge as if to attack the right wing of Smith's advancing columns. Two squadrons of the 16th were covering that flank and were ordered to charge and capture the guns. Major Rowland Smyth was commanding the Regiment that day, the Regiment's Commanding Officer, Lieutenant-Colonel G.S. Macdowell, having been placed in command of the Cavalry Brigade. 'We had a splendid man for Commanding Officer,' wrote Sergeant Gould, 'Major Rowland Smyth. He was six feet in height and of the most commanding appearance. At the trumpet note to trot, off we went. "Now," said Major Smyth, "I am going to give the word to charge, three cheers for the Queen." There was a terrific burst of cheering in reply, and down we swept on the guns.'

The Sikh gunners worked the guns until the very last

moment, throwing themselves under them to avoid the lances as the 16th thundered past. Just beyond, partially obscured by the dust and smoke, a dense mass of infantry presented the charging horsemen with a wall of swords, shields and bayonets. There was only one thing to do, and Smyth did it. He put his horse at the Sikhs as if at a fence in the hunting field, and cleared the front line of them. He then galloped through the centre of the triangle, collecting a bayonet wound in his back as he went, and jumped out again on the far side. His men came after him.

'We had to charge a square of infantry,' says Sergeant Gould. 'At them we went, the bullets flying round like a hailstorm. Right in front of us was a big sergeant, Harry Newsome. He was mounted on a grey charger, and with a shout of "Hullo boys, here goes for death or a commission," forced his horse right over the front rank of kneeling men, bristling with bayonets. As Newsome dashed forward he leant over and grasped one of the enemy's standards, but fell from his horse pierced by nineteen bayonet wounds.

'Into the gap made by Newsome we dashed, but they made fearful havoc among us. When we got to the other side of the square our troop had lost both lieutenants, the cornet, troop sergeant-major, and two sergeants. I was the only sergeant left. Some of the men shouted, "Bill, you've got command, they're all down". Back we went through the disorganized square, the Sikhs peppering us in all directions. We retired to our own line. As we passed the General, he shouted, "Well done 16th. You have covered yourselves with glory." Then noticing that no officers were with C troop, Sir H. Smith enquired, "Where are your officers?" "All down," I replied. "Then," said the General, "go and join the left wing under Major Bere." '

The 16th paid a heavy price in dead and wounded men and horses. In Smyth's case a bayonet had entered his body just below the waist and then broken off in the wound. Faint with pain and loss of blood, Smyth had nevertheless led the return charge through the broken Sikh ranks, and only after this allowed himself to be taken to the rear. He was in agony, the bayonet having pushed part of his tunic and sword belt into his stomach, but he refused to allow his wound to be dressed until

Major J. Rowland
Smyth commanded
the 16th Lancers at
the Battle of Aliwal

Charge of the 16th Lancers at Aliwal

the rest of his wounded had received attention. Six weeks later he was back in the saddle and he lived to a ripe old age.

The charge of the 16th Lancers proved to be the culminating point of the battle. The Sikhs began to fall back, the village of Bundri was taken by a bayonet charge, and the British guns were brought forward on to the ridge where formerly the Sikh guns had been emplaced. A confused mass of Sikhs fell back towards the river bank, hemmed in by Cureton's cavalry on the flanks and blasted by Smith's guns at point-blank range. Only a bruised and bleeding rabble managed to reach the far bank, leaving behind all their guns and 3,000 dead and dying. They had fought well. Waterloo veterans said they had fought as bravely as the French. British losses were relatively light but the 16th were not so fortunate. Two officers and fifty-six NCOs and soldiers had been killed; six officers and seventy-seven NCOs and soldiers had been wounded, of whom thirty died later. Seventy-seven horses were killed, thirty-five were wounded, and seventy-three were reported missing. The Regiment had mustered about 500 men and horses at the outset of the battle. By the time the Cease Fire was sounded it had lost more than a third of its strength. Little wonder that Captain Pearson, who led a squadron, wrote home to say, 'The 16th Lancers had suffered most severely, much more so than in any battle in the Peninsula or at Waterloo.'

Aliwal was only one of several bloody battles fought during the First and Second Sikh Wars but it made Sir Harry Smith's reputation as a general. The Duke of Wellington paid him a heartwarming tribute in the House of Lords while Sir Robert Peel moved the motion in his honour in the House of Commons. Public houses were named after the 'Hero of Aliwal' and the charge of the 16th Lancers captured the public imagination in the same fashion as the Charge of the Light Brigade at Balaclava was to do some years later. In South Africa, where Sir Harry Smith went on to command the forces, a railway junction was named Aliwal, and two small townships became Ladysmith and Harrismith.

Today Aliwal is only one, and only a very unimportant one, in the long string of Battle Honours acquired by the British

...e service following the 16th Lancers' return to England after the Battle of Aliwal. Their duties ...ded an escort for Queen Victoria during her State Visit to Manchester in 1851

Army in the course of the conquest of India. Now it is all but forgotten. However, on the battlefield itself there stands a crumbling obelisk to mark the spot where Sir Harry Smith stood to watch the progress of the battle; it is inscribed 'Aliwal, 28th January, 1846.'* The Regiment still faithfully commemorates the battle each year on its anniversary. And outside the Regimental Quarterguard, where a gong hangs inscribed with the Regiment's Battle Honours from a triangle of lances, the red and white lance pennons are still carefully starched and crimped. This is to commemorate the fact that when the battle was over the lance pennons were so encrusted with blood that they appeared to have been starched and crimped. This is part, however small, of the traditions of the British Army, and it has been faithfully maintained for the past 150 years.

* It is interesting that Fortescue in his *History of the British Army* gives 29 January as the date of the Battle of Aliwal. However, Graham's History of the Regiment gives 28 January as the date, as does the Marquess of Anglesey in his *History of the British Cavalry* and Hugh Cook in his *The Sikh Wars 1845-1849*. Certainly the Regiment has always celebrated Aliwal Day on 28 January and the annual Aliwal Dinner was always held on that date by the survivors of the charge. The author visited the battlefield on 28 January, 1968. It cannot have changed much in 122 years but one would have needed telescopic eyesight to see the backdrop of the snow-covered Himalayas so eloquently described by Corporal (later Captain) Cowtan. Poetic licence? This visit is described in the 1968 edition of *Scarlet & Green*.

CHAPTER SEVEN

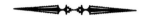

Goughie

IN 1858 the cavalry of the line were augmented by two regiments. The reason for this augmentation has never been easy to establish, nor why, of the two regiments re-formed, one should have been the old 5th Royal Irish Dragoons, under a new title, the 5th Royal Irish Lancers. The other regiment was the 18th Light Dragoons (Hussars). The General Order promulgating this increase in the number of cavalry regiments was dated 9 January, 1858:

'His Royal Highness the General Commanding-in-Chief [the Duke of Cambridge] has much pleasure in communicating to the Army the Queen's command to cancel the Adjutant General's letter dated the 8 April, 1799, announcing the Royal determination of His late Majesty George the Third, to disband the 5th Royal Irish Dragoons; that letter is cancelled accordingly.

'The Queen commands that the 5th Royal Irish Dragoons be restored to its place among the Cavalry Regiments of the line, and His Royal Highness feels assured that this mark of Her Majesty's grace and favour will be appreciated, and that the 5th Royal Irish Dragoons will emulate other regiments in discipline and loyalty, and vie with them in promoting the glory of the British Arms.'

On their re-formation the 5th Dragoons were ordered to be clothed and equipped as Lancers with the title the 5th or Royal Irish (Light) Dragoons (Lancers), soon to be shortened to the

5th Royal Irish Lancers drum banner

5th (Royal Irish) Lancers. They were authorized to assume the motto *'Quis Separabit,'* together with the Harp and Crown formerly borne on the Standard and appointments of the 5th Royal Irish Dragoons. Their strength was fixed at 660 officers and soldiers. They were to be formed at Newbridge in Ireland and Major-General Sir James Chatterton was appointed the first Colonel of the Regiment. Special inducements were offered to persuade men to enlist, including a double bounty, and both regiments were formed remarkably quickly. However, 500 recruits deserted from both regiments within a year, although the figure had fallen to less than 120 in the following year. In 1863 the 5th Lancers were posted overseas to India, their station Cawnpore. 513 other ranks embarked, accompanied by sixty-nine wives and ninety-three children! The wives left behind were left to fend for themselves, other than the cost of a ticket to the place where they intended to stay. There was at that time virtually no provision for marriage in the army.

The 5th Lancers soon settled down in India and acquired for themselves a considerable reputation. As might have been expected of a regiment with a large number of Anglo-Irish officers, they showed to advantage on the polo field and are credited with introducing into Britain the Indian cavalry sport of tent-pegging with lances, on their return home after 10 years in the 'Shiny'. They also developed a trick ride that was much

The 16th Lancers Polo Team 1880. Left to right (standing) Lieutenant F. G. Blair, Captain J. M. Babington (seated), Captain H. R. L. Howard, Lieutenant W. H. Wyndham Quinn, Lieutenant J. G. A. Baird

in demand at gymkhanas and Horse Shows. They were rather a happy-go-lucky crowd of officers, and far less 'sticky' than certain cavalry regiments. Like their future partners, the 16th Lancers, they got on with everyone. In 1884 they sent two squadrons to Suakin on the Red Sea where a force was being assembled under General Graham to relieve Gordon in Khartoum. They took part in some very tough operations against Osman Digna, one of the Mahdi's more capable lieutenants, who headed the Hadendoa tribesmen of the Red Sea Hills (Kipling's 'Fuzzy Wuzzies'). The 5th Lancers were involved in the actions at Hasheen and Tamai where the tribesmen very nearly broke into a British square. The 5th Lancers lost their Commanding Officer and eight soldiers killed. The Regiment was granted the Battle Honour Suakin, 1885, its first since Malplaquet in 1709.

There were twenty-eight cavalry regiments in the British Army from 1858 until the end of the First World War, and additionally three regiments of Household Cavalry, 1st & 2nd Life Guards and Royal Horse Guards (The Blues). The principal overseas garrisons were India, South Africa and, after 1882,

A soldier on the Gordon Relief Expedition – a silver statuette

he three officers of the 16th Lancers detachment of the force assembled to relieve eneral Gordon. 5th Lancers provided two squadrons

Egypt. Never less than six, and seldom more than nine regiments were in India where a regiment served from nine to eleven years before returning home. The 'Heavies,' i.e. the Household Cavalry, Dragoon Guards and Dragoons, did not serve in India.* Although the soldier was in most respects much better off in India than at home, the mortality rate was alarming. In 1889, when the 5th Lancers returned to India after a long spell at home, they lost fifty rank and file from enteric fever within the first six months. There was also a very high rate of venereal disease in India and a great deal too much drunkenness. The cause was of course the great amount of spare time owing to the climate.

Until almost the end of the nineteenth century every regiment serving overseas maintained a depot at home, usually at Canterbury, but sometimes at Colchester. Recruits were trained under the supervision of two regimental officers posted there for the purpose. Recruits were enlisted at eighteen but were not allowed to be posted to India until they were twenty. Much of the time spent at the depot was wasted, the staff being insufficient to train them properly, and there were even instances of recruits being posted overseas who still could not ride properly. In 1890, for example, there were only 388 horses available for riding school work but there were 1,248 rank and file to ride them. In 1897 Canterbury and other depots were abolished and from then onwards regiments on the home establishment were called upon to supply drafts for regiments overseas.

In 1889 the 16th Lancers were in Aldershot. There they were joined by a nineteen-year-old subaltern, Hubert de la Poer Gough, from the Anglo-Irish family of that name. His father and his uncle had both won the VC in the Indian Mutiny, and Hubert's brother also won the VC; it was a remarkable family record. Hubert Gough was always known as 'Goughie' in the 16th Lancers, becoming one of the Regiment's brightest stars. In his memoirs he described what life was like in a British cavalry regiment eighty years ago:

'The 16th Lancers were an extravagant regiment, but not more so than many other cavalry regiments. Life was certainly

* This changed after 1889. Both the 1st King's Dragoon Guards and the 2nd Dragoon Guards had already served in India in the 1880s.

Russian silver and enamel tankard presented to the 16th Lancers in 1889 after
the Regiment provided an escort for the Czarina during a visit to London.
The Kremlin buildings are embossed around the vessel

gay.... We had all sorts of dress for different occasions — full
dress scarlet tunics, gold lines and girdle; stable jackets, which
were short jackets gold-laced and hooked up to the throat;
braided blue frock-coats, braided patrol jackets, and, of course,
mess dress.... As we were scarlet lancers, when our NCOs and
men walked out they wore a scarlet monkey-jacket, hooked up
to the top of the throat, and a stiff blue collar.... The overalls
were strapped under the instep, of dark blue, and with two
yellow stripes down the sides, and a pair of well-burnished steel
spurs in the heel of their Wellington boots. We all wore, when
in undress, a small circular forage cap with no peak.... and we
carried a small cane.

'We were all very proud of the Regiment and on very good
terms with each other.... There was no squadron organization
when I joined, all administration was by troops, of which there

were eight in each cavalry regiment. We were expected to know our drill perfectly and to be letter perfect in all words of command... practically all our mounted work consisted of drill movement.... It was strange indeed that the British cavalry — and indeed all arms of the British Army, who had learnt so much under Moore and Wellington in the Peninsular War — had neglected and forgotten all they then knew and practised. After Waterloo the British Army went back to the days of Frederick the Great.'*

Military thought certainly stagnated for the greater part of Queen Victoria's reign. Part of the problem was due to the fact that officers remained too long in command of regiments, sometimes for as long as seven to ten years. Also, in far too many instances, they attained command when too old. In 1876, for example, the Royal Commission recommended that all general officers should retire at the age of seventy! But there were some important reforms, such as the abolition of flogging in peacetime in 1868; the Duke of Wellington had always supported flogging when he was Commander-in-Chief. In actual fact it was never much practised in cavalry regiments during the nineteenth century; on the whole they recruited a better type of man, and were more popular than the infantry regiments. In 1883 the 5th Lancers had thirteen soldiers out of its strength of 469 all ranks who had 1st class certificates of education, which placed it 'by a long way at the top of the list of cavalry regiments'.**

Another reform was the abolition of Purchase in 1871. This was the buying and selling of officers' commissions. There was a 'regulation price' for each rank up to Lieutenant-Colonel, viz: Cornet £450, Lieutenant £700, Captain £1,800, Major £3,200, Lieutenant-Colonel £4,500. Additionally there were over-regulation prices for each commission other than Cornet. In the 5th Lancers these were: £750 for Lieutenant, £4,200 for a Troop (Captain), £6,400 for Major, and £8,700 for Lieutenant-Colonel, i.e. command of the regiment. The 16th Lancers were more expensive. Supporters of the system claimed that it ensured that officers had a vested interest in the regiment; when they retired they had lump sums to fall back

* *Soldiering On* by General Sir Hubert Gough (Arthur Barker, 1954)

** *A History of the British Cavalry* by the Marquess of Anglesey (Leo Cooper, 1982, Vol 3, p 72).

on at a time when pensions were derisory. On promotion to Major-General, however, the individual concerned lost his entire stake, the vacancy left in the regimental list being filled without purchase.

Although the cavalry thought a great deal of themselves, officers and soldiers alike, this feeling was not shared by the rest of the Army, who regarded them as 'show-offs'. In the garrison towns there were regular fights after the pubs had closed between the cavalry and the infantry or gunners. In most of the European armies cavalry officers were considered to be wild, extravagant blockheads. Winston Churchill was scathing about the intellectual shortcomings of his brother officers in the 4th Hussars. So was Osbert Sitwell of an officer in the 11th Hussars who was so stupid that 'even his brother officers noticed it'. Writing of his early years in the 16th Lancers, Hubert Gough has said: 'I liked my brother officers personally, from the colonel to my fellow subalterns, but I could not feel much respect for them as serious soldiers. Indeed none of them were serious soldiers, if by that is meant students of tactics and strategy.'

It required the Boer War to shake the British Army out of its complacency. Until then fighting colonial wars against enemies without modern weapons had been all too easy. But there were

The Riding School – "Halt!" At the turn of the century

nevertheless some officers and NCOs who did take their profession seriously. Hubert Gough was one of them; he actually passed into the Staff College, almost unheard of in the 16th Lancers at that time, and even went back there later on the Directing Staff. Another dedicated soldier was a young man from Lincolnshire, son of the village tailor and postmaster, who enlisted in the 16th Lancers as a recruit on 13 November, 1877. By sheer force of character and determination he rose through the ranks to obtain a commission, ending his service as a Field-Marshal and a Baronet of the United Kingdom. He was William Robertson.

Robertson, who had entered domestic service aged thirteen, joined the Army to his mother's great distress. She said the army was a refuge 'for all idle people. I would rather bury you than see you in a red coat.' However, her son rose rapidly to be a Troop-Sergeant-Major in 1885. He never missed an opportunity to increase his military knowledge and received the support of several of his officers. They persuaded him to apply for a commission and in 1888 he was gazetted a 2nd-Lieutenant in the 3rd Dragoon Guards. He chose that regiment because it was serving in India where expenses were less than at home. Robertson augmented his pay by the study of languages, for which he had a gift, which led eventually to a posting to the Intelligence Department at Army Headquarters. In 1894 he married a general's daughter, which helped him socially in an overwhelmingly snobbish society, and in 1897 he succeeded

William Robertson as a Corporal in 1879, later to become Field-Marshal Sir William Robertson

Goughie

Lieutenant-Colonel James
Babinbton pigsticking, India
1894. An oil painting by
Lieutenant Clive Dixon, 16th
Lancers

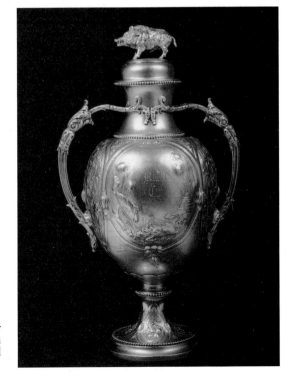

The coveted Kadir Cup for
pigsticking, won by Captain
M.P.R. Oakes, 5th Royal
Irish Lancers in 1891

in entering the Staff College, of which he was to be Commandant from 1910-13. Always gruff and plain-spoken, he was, as one of his contemporaries put it, 'as solid as an English oak, and just as endurable'.

The 16th and the 5th Lancers were both in India in 1897. The 5th Lancers were commanded by Lieutenant-Colonel 'Jabber' Scott Chisolme who, at the end of the training season, wheeled his regiment into line and ordered his trumpeter to sound the 'Officers' Call'. After they had galloped out to him, he told them: 'Gentlemen, I have called you out to look at such a regiment of cavalry as you are unlikely ever to see again. Turn about and look at the Regiment.' A junior officer then present later recorded: 'I knew the Colonel was right as with pride we gazed on that long line of five strong squadrons standing motionless under the Indian sun, not a horse out of place.' It is clear that the 5th Lancers had established an excellent reputation for themselves in India and when they left for South Africa in 1898 the GOC Bengal reported: 'A first-rate regiment in first-rate order. I never saw a better. It is in every way fit for active service. I am sorry to lose it from my command.'

They were soon to be put to the test of war. When war with the Boers broke out in October, 1899, the 5th Lancers were at Maritzburg in Natal. Scott Chisolme had left them and went

Final charge of the 5th Lancers at Elandslaagte

The "Chevril" factory. "Emaciated troop horse was issued in a disguised and more palatable form to the troops and inhabitants of Ladysmith during the siege. 'Chevril', chevril paste, sausage meat, and even calves-foot jelly, were all turned out from the same raw material". From Captain Clive Dixon's (16th Lancers) book of drawings *The Leaguer of Ladysmith*

on to raise the Imperial Light Horse. On 21 October, 1899, the 5th Lancers, together with the Imperial Light Horse, charged and scattered the Boers at Elandslaagte, one of the few occasions during the South African War when old-style *l'arme blanche* tactics came into their own. Unfortunately Scott Chisolme was killed after dismounting to bind up the leg of a wounded trooper. This cast a gloom over what had been a very successful action where the 5th Lancers were concerned. Two weeks later they found themselves besieged in Ladysmith along with the rest of Natal Force under Sir George White, VC.

The siege lasted four months and rations were short. 'For dinner today,' wrote one diarist on 2 February, 1900, 'we had chevril (a horse bovril) and the haunch of a mule. No doubt one could manage it if the meat and soup were good instead of being tainted by the hot weather and the plague of flies.'

Ladysmith was relieved by a mounted column of the Imperial Light Horse and Natal Carabineers; they were all volunteers. Hubert Gough, who was serving with this force, was the first officer into Ladysmith. There he was greeted by Sir George White, pale and drawn but otherwise in perfect control of his emotions. 'Hullo, Hubert, how are you?' was all he said. Later Gough met his brother John, who had served throughout the siege with the Rifle Brigade. He too greeted his brother in matter-of-fact fashion. 'How fat you've got,' were his welcoming words! Another member of the relief column was a young officer who had served formerly with the 4th Hussars, and who had charged at Omdurman with the 21st Lancers. He was Winston Spencer Churchill who was altogether too obstreperous for Hubert Gough.

The 16th Lancers did not arrive in South Africa until early in 1900; they too came from India. They took part in the Relief of Kimberley and the Battles of Paardeberg and Diamond Hill, and did not return home from South Africa until November, 1904. The 5th Lancers returned from South Africa in October, 1902. Kipling's verdict on the South African war was that the British Army had had 'no end of a lesson'. This was undeniable, but whether it did us 'no end of good' is less easy to sustain.

Prisoners captured by the 16th Lancers from de Wet's commando at Paardeberg

1e Relief of Ladysmith

The army that set out to deal with the guerrilla tactics of the Boers, mounted infantry of the highest class, still practised the tactics of the Napoleonic and Crimean wars. The cavalry still believed in the 'Charge!' which was useless if the enemy was not prepared to wait to be charged. In any case the invention of the Maxim gun, and similar type weapons, made the mounted man the most vulnerable object on the battlefield, particularly when the enemy were first-class shots. Gradually, after several humiliating experiences, tactics improved, new ideas were introduced, and out-of-date generals were removed. The result was a much improved army but a lot of good men died before the lessons were learnt. It is an interesting statistic that 518,000 horses were provided, of which 347,000 were expended, an average daily loss of 336!

In operational uniform

In Affectionate Remembrance of
JILL
A LITTLE DACHSHUND THE PROPERTY OF
CAPT. M. F. Mc TAGGART 5TH LANCERS
BORN IN SIMLA APRIL 1895 ACCOMPANIED
HER MASTER WITH THE REGIMENT TO
SOUTH AFRICA 1898 1900 & WAS ONE OF THE
FEW DOGS THAT SURVIVED THE SIEGE
OF LADYSMITH DIED IN YORK 7TH FEB. 1908

Gravestone found in York. It is now placed in the Officers' Mess garden wherever the Regiment may be

Aynsley china commemorative of the Boer War

5th Lancers Maxim Gun Team 1904

Two officers who made their name in South Africa and who later became Colonels of the 5th and 16th Lancers were Edmund Allenby and James Babington. Allenby, who commanded the 5th Lancers from 1902-05, on transfer from the 6th Inniskilling Dragoons, was undoubtedly the finest cavalry leader produced by the British Army during the First World War. James Babington, who first came to the fore in South Africa, was a 16th Lancer. He commanded the British Army in Italy towards the end of the First World War and ended his military career as a Lieutenant-General; both his son and grandson served in the Regiment after him. Allenby of course ended his career as a Field-Marshal and would have been Chief of the Imperial General Staff had his predecessor in the appointment, Sir Henry Wilson, had anything to do with it. The Prime Minister's choice was, however, Lord Cavan. Allenby became British High Commissioner in Cairo instead, dying in 1936. After cremation, his ashes were interred in Westminster Abbey, one of the only two First World War leaders so honoured; the other was Field-Marshal Lord Plumer.

The Scarlet Lancers

It used to be the fashion to make foreign rulers honorary generals or field-marshals in the British Army and appoint them honorary colonels-in-chief of British regiments or Corps. In 1905 King Alfonso XIII of Spain was appointed Colonel-in-Chief of the 16th Lancers, and within the limits of that appointment continued to take an interest in the Regiment for the rest of his life. In 1906 he was married to Princess Victoria of Battenberg in Madrid. Four officers of the Regiment were invited to attend the ceremony. A bomb was thrown at the King and his bride as they drove through the streets, and the four 16th Lancers officers walking behind the royal carriage hastened to the King's side. Fortunately neither he nor his bride was injured but the 16th Lancers insisted on escorting the King back to the Royal Palace. All four received Spanish decorations.

The Regiment's close connection with King Alfonso has resulted in two interesting traditions, one certainly founded on fact, while the other may well be apocryphal. The first tradition is that on band nights in the officers' mess the Spanish Royal Anthem is played immediately before the National Anthem.

THE BRITISH AMBASSADOR AND THE KING'S BRITISH REGIMENT TO THE RESCUE: SIR M. DE BUNSEN AND OFFICERS OF THE 16TH LANCERS ASSISTING THE QUEEN OF SPAIN TO ALIGHT FROM THE ROYAL COACH AFTER THE BOMB EXPLOSION. (From a sketch by our Special Artist-Correspondent.)

At the time of the explosion of the bomb the British Ambassador, Sir M. de Bunsen, and the officers representing the 16th Lancers, the British regiment of which King Alfonso is colonel, were quite close to the Royal coach. They hurried forward and assisted Queen Victoria to alight from the damaged vehicle, and afterwards on foot escorted the emergency carriage, in which the Royal couple continued their journey, to the palace.

893

King Alfonso's wedding 1906. A drawing from a contemporary magazine

King Alfonso XIII was appointed Colonel-in-Chief of the 16th Lancers during his State Visit to King Edward VII in 1905. Portrait in full dress uniform of the Regiment by the distinguished Spanish artist Joachim Serolla y Bastida dated 1908

The Forge and Farriers, 1904

5th Lancers at Stables, August 1907

Unless there happen to be foreigners dining, the officers do not drink the Loyal Toast nor stand for either the National nor Spanish Anthems; the origins for this are shrouded in mystery. The second tradition, or perhaps it should be described as a peculiarity, relates to the practice of wearing the cross-strap of the Sam Browne belt the wrong way round, i.e. fastening at the back rather than at the front. Tradition has it that King Alfonso was inspecting the Regiment at Tidworth, on Salisbury Plain, in the early 1920s. The Commanding Officer was Lieutenant-Colonel (later Major-General) Geoffrey Brooke, an Anglo-Irishman in the Gough tradition, a brilliant horseman, excellent writer and true *beau sabreur*. When, prior to going on parade, the King appeared in Service Dress uniform, wearing his cross-strap the wrong way round, Geoffrey Brooke felt he could hardly tell the Monarch that he was improperly dressed! He therefore told all the officers to fasten their belts in the same way as the King's, and they have been worn in this fashion ever since.

Many of these idiosyncracies in dress or mess customs owe their origins to the whims of individual commanding officers that become hallowed by time. Here is but one example of this. When battledress was introduced early in the Second World War, it was usual to tuck the trouser-ends into short canvas gaiters, and this was accepted parade dress for most regiments, including the Regiment. In 1949, however, the then Commanding Officer, Lieutenant-Colonel T. C. Williamson, who had come from the 5th Royal Inniskilling Dragoon Guards, took exception to the wearing of gaiters; instead he decreed that battledress trousers would be shaped to fit over the boot, and gaiters would no longer be worn. When the Regiment was in Catterick in 1958, it was ordered to provide a contingent to take part in the annual military parade at York Minster; the order of dress included gaiters. This caused something of a stir when first received, but it was followed by a telephone call from the Chief of Staff at Headquarters Northern Command in York. 'The order regarding gaiters does not of course apply to you,' the General told the author, who was commanding the Regiment at the time. 'The Army Commander knows it is due to a Battle Honour [he did not say which!] that your Regiment is excused wearing gaiters in battledress.'!

Hubert Gough became Commanding Officer of the 16th Lancers at the early age of 37. Not only had he attended the Staff College, which was not particularly encouraged in the 16th Lancers in those days, but he had later returned to Camberley as an instructor. Gough was clearly an officer with a future. According to Brigadier-General Beddington, another very able officer who served under him, Gough was an inspiring Commanding Officer. His interest lay in training, not the parade ground, and he was a first-class tactician. 'By the time Goughie's command was over,' wrote Beddington, 'we were just about the best-trained cavalry regiment in the Army.' Gough was equally keen on sport and insisted his officers kept fit. He was a good disciplinarian but he had a mind of his own and was not afraid of speaking out when he disagreed with his superior officers. He was brave both physically and morally, the latter being the rarer quality among officers with ambition.

It was this moral courage that got Gough into trouble in March, 1914. At the time he was commanding the 3rd Cavalry Brigade, consisting of the 16th Lancers, 4th Hussars and 5th Lancers, at The Curragh in Ireland. Ireland was in the throes of a crisis, caused on this occasion by the refusal of the Protestants in Ulster to accept Home Rule. Led by Sir Edward Carson and James Craig, they had said they would rather fight than accept, but Gough has said that the majority of British soldiers serving in Ireland were not much interested in what they conceived to be a political quarrel. There have been many books published on what has come to be known as the 'Curragh Mutiny', which it certainly was not, but Beddington's account is as good as any, and probably less opinionated than most. He had been lecturing to the Staffordshire Yeomanry and was returning to the 16th Lancers when he was cheered by the porters on Crewe railway station. He enquired the reason for his popularity, to be told: 'Your regiment refused to go and fight in Ulster; here's luck to you all.' Beddington goes on to say:

'I called at Barracks on my way home and found that on the Friday (20 March) General Paget, G.O.C. Irish Command, had

General Sir Hubert Gough. *Robert Hunt Library*

sent for Goughie and others and told them that they and their officers had the choice of resigning their Commissions (in fact being dismissed) or of most probably engaging in active operations against Ulster. Any officer with an Ulster domicile would be allowed to disappear. Paget required an answer that evening. Goughie saw the 5th Lancers in Dublin that morning, and the other officers of the Brigade in our mess that afternoon, and the vast majority (*all* in the 16th Lancers) decided for resignation or dismissal; and Gough reported to Dublin that evening that of the officers on duty 59 out of 71 preferred dismissal. The next day General Paget went to The Curragh to see the officers of the Brigade stationed there, and endeavoured to get them to reconsider their decision; there was some weakening but very little. Gough, and the colonels of 5th and 16th Lancers were ordered to report to the War Office the next day, Sunday.'

Seldom in the long and sorry history of England's relationship with Ireland has anything been worse handled. Gough and his officers acted throughout with calmness and discipline, but the business was grossly mishandled at the highest level. It became a political sensation and led to the resignations of the Secretary of State for War (Seely) and the C.I.G.S. (French). General Paget, who was responsible for the nonsense in the first place, was retired shortly afterwards. It could have damaged the careers of many of the officers involved, not least Gough's, but war broke out with Germany on 4 August, 1914, and men's minds were focused on sterner issues than the evidently insoluble problems of Ireland.

Gough took his Cavalry Brigade to France on 15 August. One week later one of his batteries – E Battery R.H.A. – fired the first British shell in a war in which millions and millions of artillery shells were expended. The 16th Lancers had their first casualties on 22 August when Lieutenant Tempest-Hicks' Troop came upon German infantry when reconnoitring at Peronnes, near Mons. The enemy were concealed among recently cut wheat stooks. They opened fire on the 16th Lancers' patrol which immediately charged them. Tempest-Hicks' horse was shot under him; two other horses were killed and a soldier was wounded. However, covering fire from E Battery enabled the patrol to withdraw safely. Gough wrote of Tempest-Hicks that he was 'one of our most gallant young officers'. He survived throughout most of the war, mostly in the trenches, but was killed by a chance shell well behind the front lines shortly before the Armistice.

There was little opportunity for the cavalry, *l'arme blanche*, to distinguish themselves on the Western Front during the First World War. It was essentially an infantryman's and artilleryman's war — of trenches and barbed wire, mud and blood, raid and counter-raid, massed attacks on foot, bombardment and counter-bombardment. Men lived in holes in the ground like animals, often up to their waists in mud, enduring every extreme of climate. Only farther afield, on the Eastern Front, or in Palestine, was mobile warfare possible. Although the cavalry generals, like Sir Douglas Haig, believed the moment would come for the 'cavalry to go through', that moment never seemed to arrive. The 16th and 5th Lancers covered the retreat of their infantry from Mons — the 5th Lancers were the last British cavalry regiment to withdraw from Mons — a retreat that was as confusing as it was exhausting; men fell asleep in the saddle. Two troops of the 5th Lancers were cut off and surrounded. They put up such a stout resistance before surrendering that almost every man was wounded. One trooper, Kay, became separated from the rest. He was called upon to surrender but took up position in a carriage where he defended himself until killed by a volley. The local villagers declared he killed six German officers before he was killed himself; his bravery so impressed them that they buried him where he fell and erected a cross above his grave.

/o drawings from the 16th Lancers Christmas card, 1916

By the middle of October, 1914, the German advance on Paris had been halted. The front had begun to stabilize and trench warfare began from the Channel coast right across France to Switzerland. The 16th Lancers went into the trenches as infantry on 20 October near Armentières; the 5th Lancers soon followed them. The latter, with an Indian regiment in support, were almost decimated at Hollebeeke by artillery bombardment. All the British officers in the Indian regiment were either killed or wounded, and the Indian soldiers began to give ground. Trooper Colgreave, who spoke Hindustani, succeeded in rallying some of them and led them back to the trenches where they fought bravely. He was awarded the Distinguished Conduct Medal (DCM).

During the winter months of 1914-15 the old regular army was bled white defending the notorious Ypres Salient. Both the 16th and 5th Lancers played their part in holding the line during some of the most bitter fighting of the entire war. Trenches were waterlogged and for much of the time men fought up to their knees in water. Even the so-called 'rest areas' were within reach of the enemy guns but there were nevertheless games of football, mounted sports and concerts of a kind. Major Barry of the 5th Lancers brought several couple of foxhounds back with him from leave. At first the French refused to release them, saying no sport of any kind was permitted during wartime, but by fair means or foul the hounds eventually reached the Regiment. For several weeks they provided good sport before the French got to hear of them and insisted that hunting had to stop.

Many writers have managed to give the impression that for much of the war in France and Flanders the cavalry hung about in the rear areas waiting for the opportunity to go through. It was an opportunity that never came until the very end. But it is not a true picture. Both the 16th and the 5th Lancers spent much of their time in the trenches as infantry. The 5th Lancers particularly distinguished themselves in the defence of Guillemont Farm in June, 1917. Although the trenches were well sited and deep, there was only one dugout in which to take refuge when the Germans opened their bombardment. It

5th Lancers' re-entry into Mons, November, 1918. Oil painting by R. Caton Woodville

The charge of the 16th Lancers at Aliwal. Painting by Michael Angelo Hayes. *National Army Museum*

16th Lancers. Oil painting by John Ferneley Jnr.

A scout car patrol with UNFICYP (United Nations Force in Cyprus). *Franklin Watts.*
See Chapter Thirteen

seemed at one time as if the position would have to be abandoned but the 5th Lancers hung on. They lost sixteen men killed and twenty-four wounded in this action, and were awarded for their gallantry two MCs, one DCM and four MMs. In December of the same year a squadron of the 5th Lancers distinguished itself in the defence of Bourlon Wood. It was during this action that Private George Clare won a posthumous VC.

He was a stretcher-bearer who dressed wounds and helped to evacuate wounded men under intense machine-gun and artillery fire. 'At one stage, when all the garrison of a detached post, which was lying out in the open... had become casualties, he crossed the intervening space, which was continuously swept by heavy rifle and machine-gun fire, and, having dressed all the cases, manned the post single-handed until a relief could be sent.' He then carried a badly wounded man through intense fire to the dressing station, where he learnt that the enemy was using gas shells in the vicinity and the prevailing wind might

Resting horses in a French village. *Robert Hunt Library*

blow the gas over the 5th Lancers' position. On his own initiative he personally warned every company post of the danger, all the time under heavy fire. This very brave man was eventually killed and the VC awarded to him was the first to be given to a member of the 2nd Cavalry Division.

Although the cavalry's role was restricted, there were a few instances when the horse came into its own. One such occasion was the charge of Lord Strathcona's Horse (a Canadian regiment) at Moreuil Wood on 30 March, 1918, when the Canadians counter-attacked the German advance and stopped them from breaking through. The Canadians suffered heavy casualties, as might be expected from horsed soldiers attacking infantry, but at a critical moment the 3rd Cavalry Brigade came to their support. In the lead were the 16th Lancers, commanded

Silver statuette depicting a soldier of the First World War

l painting by G.D. Rowlandson after de Neuville

by Lieutenant-Colonel Geoffrey Brooke. He had been Brigade Major of the Canadian Cavalry Brigade before his appointment to command the 16th Lancers and the Canadians regarded him as one of themselves. Imperturbable under fire, Brooke could see that enemy reinforcements were infiltrating into Moreuil Wood and extending their right flank outside the wood. He led the 16th Lancers in a dismounted attack against the southern corner of the wood, while the 4th Hussars operated mounted on his flank. The attack was successful but there was bitter fighting in the wood.

Brooke had the reputation of being completely unmoved by enemy fire, but not every soldier could hope to be quite so stout-hearted. In his account of the attack, Brooke wrote:

'I had just passed the word down the line to advance when a soldier, who had temporarily lost his nerve, started to run back. I had a large pair of wirecutters which I hurled at him and hit him on the knee. This may have restored his equanimity as he then carried on, or more likely, it may have been due to the remark of an old soldier seeing the German machine guns ripping up the grass in front of us. "God," he said, "it reminds me of old Nobby cutting up the billiard table." ' Just before Moreuil Wood a composite cavalry regiment was formed from the 3rd Cavalry Brigade and placed under the command of Lieutenant-Colonel Geoffrey Brooke. Its task was to hold the important Porquericourt Ridge which the Germans were threatening to occupy. Brooke led his Regiment at full gallop to reach the ridge before the advancing Germans but arrived only to find that he had been beaten to it. All but the highest part of the ridge had been seized by the enemy. The highest point was No-Man's-Land and the British were the first to get there, holding it against every German attempt to dislodge them. In Harvey's *History of the 5th Lancers*, he writes: 'I am told that there has been no more splendid sight in this war than this wild rush of cavalry across fields and through villages to gain the coveted ground.'

Such opportunities were, however, rare. The cavalry may have kept their horses but when it came to battle more likely than not they would have to fight on foot. They may not have suffered as severely as the infantry, but the 16th Lancers nevertheless lost twenty-two officers and 157 NCOs and soldiers killed or died; the 5th Lancers lost eight officers and ninety-eight NCOs and soldiers. Both Regiments served throughout in France and Flanders, for most of the time in the 2nd Cavalry Division. They were both at Ypres where the old Regular Army was bled to death. They were at St Julien and Bellewaarde in 1915, in reserve for the Somme in 1916, the Scarpe and Cambrai in 1917, and at Amiens in August, 1918, when the Germans first began to crack. Both the 16th and the 5th took part in the 'Pursuit to Mons', retracing the steps they had taken four years previously; and in November, 1918, the 5th Lancers, attached to the Canadian Corps were the first British troops to re-enter Mons, just as they had been the last to leave in August, 1914. Few who had ridden out of Mons with the Regiment were with it when they rode back down the cobbled streets.

The 'Scarlet Lancer' to reach the greatest heights during the Great War was Goughie, and when his luck ran out, and no General can hope to succeed without luck, he fell the farthest. He was a Brigadier-General in 1914 at the age of 44; he was a Major-General by the following April as GOC 7th Infantry Division; by July he was a Lieutenant-General as GOC 1st Corps. In May, 1916, while not yet 46, he was selected to command the Reserve Army, which subsequently became Fifth Army. It was remarkably quick advancement when it is realized that Montgomery was 55 when he took command of Eighth Army, and Slim was 53 when appointed to command XIV Army. Gough's rapid rise was not approved by everyone; he was said to be over-ambitious and ruthless, as well as a poor picker of men. Cavalry generals were no more popular in the Great War than they had been in the Peninsula, nor were they to be in the Western Desert during the Second World War. When the Fifth Army had to bear the brunt of the great German March offensive in 1918, it was forced to give ground. It was under strength for the length of front it was expected to defend. Gough had been pleading for reinforcements for months past but his pleas were ignored at General Headquarters. Gough fought an able delaying action which in the end managed to stem the German tide, but at a heavy cost. As is the British way in warfare, there had to be a scapegoat — to save the necks of the Prime Minister (Lloyd George) and his ministers. People forgot, conveniently of course, that Gough had never ceased to warn that he had insufficient troops to hold his sector of the line securely. There was inevitably a *sauve qui peut* in Parliament, the War Office, GHQ in France, and everywhere else where ambitious officers and politicians were concerned about their future careers. Gough was sacked without hope of redress on 28 March, 1918, and, being the man he was, he forebore to add to the terrible responsibilities of his Commander-in-Chief, Sir Douglas Haig. Gough went quietly. He returned home to a succession of interesting, but minor, appointments – one of them in Russia resisting the Bolshevik revolution – and retired from the army in 1920. Throughout the rest of his long life (he died in 1963, aged 93) he never ceased to campaign for the removal of the slur on the record of the Fifth Army, but he never said a word on his own behalf. He was fully exonerated when the Official History was published in 1936, and in 1937, to his Regiment's delight, he was made GCB. He was Colonel of the Regiment from 1936-43 and is certainly one of the most outstanding and most attractive officers ever to have served in the 'Scarlet Lancers'.*

It is always easy to forget that the end of a long war leaves the Army with as many troubles as it had during the fighting of it, probably more. This was as true in the aftermath of the First World War as proved to be the case after the Second World War. Most of the pre-war regular officers and long-service NCOs have been killed, or promoted out of the regiment, or have seized the first opportunity to leave the service for 'greener pastures'. It happened in 1919-20, as it was to do again in 1946-47. The 16th Lancers were sent to Syria in June, 1919, to help 'police' what had previously been a

---

* Field-Marshal Viscount Allenby was Colonel of the Regiment until his death in 1936 when he was succeeded by General Sir Hubert Gough. Gough, who was much younger than Allenby, was at one time Allenby's principal staff officer. Although admiring his talents and his drive, Gough thought Allenby unduly fussy and hidebound, but together they must have made a formidable combination!

province of the Ottoman Empire. It is clear that they were not in good shape. In October, 1919, the GOC in Egypt, Lt-General Sir W.N. Congreve, wrote to the CIGS, Field-Marshal Sir Henry Wilson: '16 Lancers are in a bad way, very bad. Only 5 officers at duty, a large number of men sick (from malaria), quite out of proportion to other regiments, and their horses very bad for want of proper attention. They are full of recruits and had no officers to look after them. I hope some are on the way for the regiment now is not fit for service of any sort.' As always the remedy lay in the hands of the officers, particularly the Commanding Officer, Lieutenant-Colonel H.C.L. Howard, a pre-war 16th Lancer. By the time the 16th Lancers moved to India in January, 1921, Howard had pulled the Regiment together and soon established for it a reputation in India that was second to none. He deserves to be remembered.*

* Colonel Cecil Howard commanded the Regiment from 1921-25, and was Colonel of the Regiment from 1943-50. He was the first Commanding Officer of the amalgamated 16th/5th Lancers.

CHAPTER EIGHT

16th/5th Lancers

THE Great War of 1914-18 saw the end of the old order of things. Nothing would be quite the same again. Of the thirty-seven officers on the regimental list in 1914, only fifteen were still serving with the 16th Lancers in 1919. It was the same kind of situation with the 5th Lancers when they embarked for India at the end of that year. The minds of some cavalry officers may still have been set in the pre-war mould but the entire nature of war had changed. As is usual, however, at the end of all Britain's wars, the first requirement was to reduce the size of the Army, and thereafter to spend as little money on it as could possibly be contrived.* Regardless of the commitments imposed by government policy, such as the overseas garrisons for example, the strength of the Army must be reduced as quickly as possible. This policy of retrenchment directly affected the 5th Lancers who were stationed at Peshawar on the North-West Frontier of India in March, 1921.

Their Commanding Officer, Lieutenant-Colonel H.A. Cape, was attending a race meeting when an acquaintance inquired whether Cape had seen the latest Reuter telegrams posted in the Peshawar Club. 'They are going to disband your regiment,' he added. The astonished Cape rushed off to read the telegrams, only to discover that the 5th Lancers were once again to be struck out of the Army List. It was at least two months before official notification was received from the War Office, and it then transpired that the 5th Lancers were offered the choice

* As is happening as this is being written in July, 1991!

Within the illustration:

"Handsome is as handsome does"
a Spanish Cob that distinguished itself
in jumping.

Apparently prizes can be won
without adopting weird positions.
(The Riding Establishment, Weedon

Lt.-Col. Geoffrey Brooke winner of the King's Cup.
This is the first year it has been
won by England.

A bolt!

The Italian
Seat

Lieutenant-Colonel Geoffrey Brooke (later Major-General) on his gelding, Combined Training, winner of King George V Cup 1921. Combined Training came to the 16th Lancers as an officer's charger; he was also a brilliant hunter and polo pony. He carried Geoffrey Brooke throughout World War I and was wounded by a single bullet which penetrated his neck. Brooke said "He shook his head with annoyance but apparently nothing vital was struck and in a few days he was well again". Painting by Lionel Edwards

either of conversion to a battalion of the Royal Tank Corps or disbandment. They unanimously chose disbandment.

They were not the only regiment chosen to be axed and some had powerful supporters in Parliament, not least the 5th Lancers. A bitter battle was fought to persuade the Army Council to rescind the decision, but the generals stood firm. Officers and soldiers left the service or were cross-posted to other regiments. By the end of 1921 the 5th Lancers had ceased to exist. Then, in April, 1922, there came a change of heart. Army Order 133 announced the amalgamation of certain cavalry regiments, and among them the 16th Lancers with the 5th Lancers. The combined regiment was to be titled the 16th/5th Lancers for reasons already given in the Introduction.

It is no secret that the 'shotgun' marriage between the 16th and the 5th Lancers was an unhappy one. Although the two regiments had been brigaded together over many years, they were very different in character. A further complication was that officers and soldiers of the 5th who had joined other regiments after the disbandment of their own were most reluctant to change yet again and join the 16th/5th Lancers. In fact only five officers did so, and there were similar problems with senior NCOs and soldiers. In the end it proved necessary to draft ninety-nine 16th Lancers into D Squadron (the 5th Lancers squadron in the newly amalgamated regiment). The cap and collar badges of the 16th Lancers were retained, as also various idiosyncracies of dress, such as the scarlet tunic; the 5th Lancers had worn blue. 5th Lancer buttons replaced 16th Lancer ones, and all full-rank NCOs wore the Irish Harp on their chevrons. 5th Lancer officers continued to wear their own badges and uniform for many years, including even a different mess dress. They had their riding boots made by a different bootmaker than the 16th officers, wore khaki whipcord tunics instead of khaki barathea, and even a different pattern Sam Browne belt. These differences were ardently supported by former members of the two Regiments, the Old Comrades Associations of both Regiments insisting on remaining separate.

It required another world war to knock sense into people's heads, although it has to be said that by 1939 few of those

"The New Lancing Wheel for teaching young Lancers to lance" by W. Heath Robinson

actually serving in the Regiment could be bothered with the quarrels and arguments of 1922. Inevitably, perhaps, the stronger partner in the union, in this case the 16th Lancers, became the dominant one. Under the stresses and strains of war, even the Old Comrades Associations became fused into one. Today it is only a matter of history. No one's nose was put out of joint in March, 1959, when the bicentenary of the raising of the 16th Light Dragoons was celebrated with much pomp and circumstance; and similarly in June, 1989, when the tercentenary of the raising of the Royal Dragoons of Ireland was commemorated in due style. Both Regiments have benefited by the merging of their traditions into one.

The years between the end of the First World War and the beginning of the Second were not very happy ones for the British Army. Starved of money, short of men, ill-equipped and over-stretched, it can truly be said that it was its *esprit de corps* that enabled it to meet every challenge, and that *esprit de corps* had its roots in the regimental system of the British Army. The 16th/5th Lancers spent many of the inter-war years in Egypt and India. In 1924 they escorted Lord Allenby*, High Commissioner in Cairo, when he drove through the streets to deliver the British ultimatum to the Egyptian government after the assassination of Sir Lee Stack, Governor-General of the Sudan. It was almost certainly not the best way to reconcile the Egyptians to the occupation of their country by a foreign power, but we looked at things very differently in those days.

Much of the Regiment's time was spent in routine duties — guards, fatigues and so on — which were performed with a minute attention to detail. Much time had of course to be given to the horses, the hard work in 'Stables', together with equitation, keeping everyone very fit. The horse was a great leveller in cavalry regiments, helping to bring together officer and soldier both on and off parade. Mounted sports, show jumping, polo and steeplechasing all aroused great interest, as well as competition between cavalry regiments in the garrison. However, it has to be said that the nation as a whole had lost interest in the Army, and the Army in its turn lost touch with the nation. Training was unrealistic, far too much time was

* The Field-Marshal was joint Colonel of the Regiment at the time, along with Lieutenant-General Sir James Babington.

16th/5th Lancers

The trick riding team rehearsing their act, 1930s

16th/5th Lancers going on manoeuvres, Tidworth 1933

The 16th/5th Lancers move to India, 1937. Lieutenant Dennis Smyly supervises the loading of horses

devoted to fund-raising tattoos and the like, and officers and soldiers both tended to concentrate on sport. Although the British had invented the tank, and we had in General Fuller and Captain Liddell Hart two of the foremost exponents of mechanized warfare, far too much energy was employed in the futile controversy of the Horse versus the Tank. It may seem incredible that, after the cavalry's experience between 1914 and 1918, there were still soldiers who believed that the horse still had a place on the battlefield, but that was certainly the case. It was an emotive issue and the Treasury sided with emotion because mechanization would be so expensive. The Army Estimates in 1927 provided only £72,000 for petrol, but £607,000 for forage for horses. When Parliament was informed in 1936 of the intention to mechanize eight cavalry regiments,

Some of the first motorized vehicles issued to the Regiment

the Secretary of State for War (Duff Cooper) felt constrained to apologize for such a monstrous decision. 'It is like asking a great musical performer to throw away his violin and devote himself in future to the gramophone,' he told the House of Commons.

Had it not been for Hitler, no one can possibly guess how long it would have taken to mechanize the British Army. If it took overlong at home, it took a great deal longer overseas. We sent a complete cavalry division to Palestine in 1939/40 and did not begin to mechanize it until 1941.* The 16th/5th Lancers had to wait until 1940 before they exchanged their horses for tanks. It was stationed at Risalpur (now in Pakistan) when it held its last mounted parade, and there were many misty eyes when the last cavalry command was given, 'Make much of your horses!'. Somehow one could not work up quite the same kind of feeling for a tank! Not that there were many tanks to go round in 1940 when the Regiment arrived home from India. They were issued in May, 1940, with five Great War veterans not notable for their mechanical efficiency. The great days of *l'arme blanche* were over for ever, but by no means every 16th/5th

* The author, as adjutant of his unit and stationed in Burma, was riding a horse on training exercises in September, 1941, two months before the Japanese attacked Pearl Harbor!

Lancer could reconcile himself to the fact. Some of them never did and took themselves off to military employment more congenial than with tanks. Time was to show, however, that cavalry regiments were able to adapt themselves to the new kind of warfare, although rivalry with the Royal Tank Corps added to the spice of life. 'Donkey Wallopers' the Royal Tanks called the cavalry! There were inevitably teething troubles to start with, and the Regiment was greatly indebted to its Commanding Officer, Lieutenant-Colonel Alastair Macintyre, a fine horseman who was determined to make the 16th/5th Lancers as proficient with armour as they had been with horses.

On 12 November, 1942, the Regiment sailed from the Clyde to North Africa as part of the 6th Armoured Division. Lieutenant-Colonel Geoffrey Babington was the Commanding Officer. The first three months were spent between Teboursock and Medjez el Bab — 'Three months of rain and mud, dreary days in tank bivvies by the side of our Valentines and Crusaders. Days spent in brewing up our food into all forms of messes and

Training in Valentine tanks in UK before leaving for North Africa. *Imperial War Museum*

H.M. The King, accompanied by Lieutenant-Colonel Geoffrey Babington, inspects the Regiment in Scotland in November, 1942 before embarking for North Africa. *Imperial War Museum*

hashes, and in drying out our clothes.' The first tank casualty was a Valentine, brewed up by an enemy plane through the thickest part of its armour! The news after Christmas that the Regiment was to be issued with brand-new Shermans was enthusiastically received, but before the new tanks could be made operational orders came for action. The 16th/5th Lancers therefore fought their first battle since 1918 in the obsolescent Crusaders and Valentines.

If the first three months had been boring, there was excitement enough to follow. They were heavily involved in three major battles — Fondouk, Kournine, and the final battle for Tunis. There was also a fierce action at Bordj on 11 April, 1943, when the Regiment actually 'charged' the enemy in the old cavalry style. Bordj was unusual since the Regiment operated as a whole instead of in squadron packets. 'A great clang, and the turret was full of flames,' says one account. 'It seemed only a fraction of a second that I was struggling to open the front lid of the turret, but in that time the gunner slipped up from his seat in front of me, pushed open the lid and clambered out. It was getting unbearably hot. I remember thinking I must get out quickly before the heat took my

127

The Regiment moves up to the assembly point in Valentines south of Thala, North Africa. *Imperial War Museum*

strength. With a heave I made it and jumped from the top of the turret to the ground, rolling over and over to put my clothes out.' Although burnt, wounded and badly shocked, he lived to fight another day, but being 'brewed-up' by enemy gunfire was always the armoured soldier's nightmare.

It was fortunate that the Regiment had disposed of their Valentine and Crusader tanks in return for Shermans by the time they became heavily involved in North Africa. Their faith in the former had received a nasty jolt at Sbeitla when a German Tiger tank had knocked out three Crusaders at a range of nearly 2,500 yards. The Sherman, however, mechanically reliable with reasonable protection and a good gun, was a different proposition. It served the Regiment well for the rest of the war, and was indeed performing yeoman service in other armies 25 years or more after the war was over.

Although the German situation in Tunisia seemed hopeless, they fought fiercely until the very end and took their toll of the 16th/5th Lancers. The Regiment buried three officers and

sixty-one NCOs and troopers in North Africa, but their sacrifice
was not in vain. On 13 May, 1943, Field-Marshal Alexander was
able to make his famous signal to the Prime Minister:

'Sir, it is my duty to report that the Tunisian campaign is
over. All enemy resistance has ceased. We are masters of the
North African shores.'

So far as the 16th/5th Lancers were concerned the campaign
in North Africa had set the seal on two years of hard training
in Britain, followed by months of hard fighting as an armoured
regiment. They were now seasoned tank soldiers and part of a
well-trained team – the 26th Armoured Brigade of the 6th
Armoured Division, whose mailed first insignia they wore on
the sleeves on their battledress blouses. With them in the 26th
Armoured Brigade were the 17th/21st Lancers and the 2nd
Lothians and Border Horse (Yeomanry). There was a lull of
eight months, during which the Allies drove the enemy out of
Sicily, before the Regiment found itself in action again. This
time it was in Italy where they were to experience some of the
toughest campaigning in all the long history of the Regiment.

'During the seventeen months in which we were engaged in
Italy,' writes the Regimental Chronicler in *Scarlet and Green*,
'I do not suppose it was for more than five that we were
employed in conditions as any armoured regiment enjoys –
that is, with room to manoeuvre, in a normal countryside,
squadrons supporting squadrons, and together (perhaps with
infantry) hustling the enemy with the enormous fire power we
now have in our tanks. Instead, you would find us in vineyards
and olive groves, a tank here, and a tank there; a squadron split
up on an infantry brigade front, grappling with mud and with
minefields, with a field of view not much longer than the length
of our guns. That was Minturno. Again you might have found
us perched in our tanks on the very heights of the Apennine
Mountains looping our way up a zigzag lane hoping to force a
way through some suspected defile and to debouch behind the
enemy into the Po Valley. At another time we became infantry.
Deserting our tanks, again in the mountains, we undertook six
weeks of vigorous patrolling on our feet – and in deep snow!'

For most of the time it was tough, hard slogging from one

Milan

Venice

Po

Casola
Valsenio

Florence   Pesaro

Arezzo
Osimo

Perugia

A
p
e
n
n
i
n
e
s

Adriatic Sea

CORSICA

ROME

Liri
Melfa

Anzio   Cassino

Minturno   Garigliano

Naples

SARDINIA

Tyrrhenian
Sea

SICILY

Miles

Malta   0   100

: Sergeant J. Harvie and his tank crew, Trooper J. Hudson and Trooper K. Marland reading azines just received from home. *Imperial War Museum*

river line to the next, with rain, mud and the snow as much the enemy as the Germans. At Minturno, and the crossings of the River Garigliano, the task was to help the infantry forward. 'It was nothing like tank country: with close cultivation, high hedges, olive trees and grape vines, you could seldom see more than a few dozen yards in front. Tanks got bogged in bottomless ploughs and the enemy were sited with anti-tank guns too well placed to allow risks to be taken... Sometimes we had one squadron up with two back, and sometimes the other way round. Those that were back trained with the infantry, practised cooperation with them and generally got to know their wants and their ways. Then they went up to the line. It was a laborious and tiring business. Each squadron would stay up about ten days and then go back for a breather, a bath at the mobile showers, a trip or two into the city [Naples] to see an E.N.S.A. show and more training.'

The abortive battle for Cassino followed, when the Regiment waited eagerly for the opportunity to break through, an opportunity which never came, despite the gallant attempts by the infantry to break the German resistance. By May, 1944, the Regiment had rejoined the 6th Armoured Division, now concentrated in Italy, and the long-awaited final assault on Cassino took place. On 15 May the 16th/5th Lancers began to move forward, with very limited objectives in view of the close country and the débris of the battlefield, but they were moving in the right direction. It was on that same day that the Regiment lost its very popular and gallant Commanding Officer, Lieutenant-Colonel John Loveday, who was killed arranging an attack to consolidate the Regiment's position. Loveday's loss was deeply felt in the Regiment. He had had a miraculous escape from death in Tunisia. Captured by an enemy patrol, he and his companion were put up against a wall to be shot. Loveday managed to stand sideways to the firers and fell simultaneously with fire being opened. He was unhurt but feigned death, even when one of the Germans took off his wrist watch and went through his pockets. Later he regained his own lines. The sergeant who was with him was killed. Had Loveday lived, it is virtually certain that he would have reached high rank. He was a truly professional soldier with great leadership qualities.

The advance up the Liri Valley followed the breakthrough at Cassino. It was during this advance that A Squadron particularly distinguished itself in the capture of Piumarola on 17 May, 1944. The fight for Rome was now on. After crossing the River Melfa the country was more open and the armour came into its own. 'We went like scalded cats, advancing so swiftly we all but ran off our maps, in fact, at one place, the Intelligence Officer had to stand by the side of the road handing out maps of the country ahead to each tank as it passed.' Rome fell on 4 June but the advance continued. Perugia fell on 20 June. 'There we stayed for ten days – ten days to reorganize, rest and do repairs. It was a very pleasant town and we got ourselves some splendid billets — a girls' school. The girls, worse luck, had gone!' Perugia was the first real break since Cassino, and it was to be the last for many weeks to come. Florence was to be the next objective and it seemed to be almost round the corner — three weeks' fighting at the most. However, it turned out to be nearer eight weeks. The Germans were fighting very skilfully and there was hard fighting all the way to Florence, including a hard battle for Arezzo, which the Regiment entered on 10 July. Five days previously it had lost yet another Commanding Officer, Lieutenant-Colonel Gerald Gundry, who was wounded by a shell fragment. Lieutenant-Colonel Willy Nicholson, himself a 16th/5th Lancer, who had been commanding the 2nd Lothians and Border Horse, returned to the Regiment to command.*

first tanks of the Regiment advancing through the town of Arezzo. *Imperial War Museum*

* The 16th/5th Lancers had no less than seven commanding officers between their landing in North Africa and the end of the war in Italy. They were Lieutenant-Colonels G. Babington, W.R. Nicholson, R.M. Fanshawe, J.R.N. Loveday, G.A. Gundry, D.D.P. Smyly and G.H. Illingworth. Nicholson commanded the Regiment twice, viz; April-June, 1943 and July-September, 1944.

Sherman tank passes through the wrecked railway installations in Arezzo. *Imperial War Museum*

All replenishment and tank maintenance had to be done after dark when 'even a flicker of light was liable to attract shell fire'. As Florence grew nearer, the country became more difficult and mountainous. Squadrons became increasingly expert in employing indirect fire to harass the enemy and help the infantry forward. Booby traps and mines on the narrow mountain tracks took their toll of men and machines. Progress was slow as the Sappers repaired the numerous demolitions left by the Germans as they retreated farther into the mountains. This was the worst tank country in Italy, which is saying a great deal. There was a short break from mountaineering just before Christmas, 1944, when the Regiment was suddenly moved to Osimo near the Adriatic coast. It was rumoured that it was about to be sent to Greece, but the order was cancelled on Christmas Eve and by 14

Sergeant Orr and crew were presented with a home made Union Jack by some civilians in Arezzo, July 1944. *Imperial War Museum*

January, 1945, the 16th/5th Lancers were back in the mountains above Florence, this time as infantry. For nearly six weeks they held part of the line in the area of Casola Valsenio where the main problem was getting supplies and ammunition to the forward positions. Mules took the place of trucks and jeeps, bringing back nostalgic memories to those who had once served with the horse.

On relief they moved to Pesaro in March. There they were reissued with tanks and began training for what was to turn out to be the final offensive of the war in Italy. Working in close cooperation with the 1st Battalion of the 60th Rifles (K.R.R.C.), each tank squadron and infantry company was welded together into a compact battle group. The Eighth Army attack was launched on the evening of 9 April, 1945, but it was not until 19 April that the Regiment and the 60th Rifles moved forward.

Crossing the River Arno. *Imperial War Museum*

The Regiment, part of 6 Armoured Division, moving up to Castiglione, Italy.
Imperial War Museum

'For five days, leap-frogging in turn with the other regiments of the Brigade, we buffeted our way through the remaining enemy lines and lashed into his attempts to reorganize himself in front of the Po. We had wonderful support from the 60th, from our guns and from the air, and we were able as a Brigade so to bounce our way over the various river and canal lines that the enemy might otherwise have made good, that he was utterly beaten and disorganized south of the River Po and was forced to abandon practically all his equipment and vehicles'.

It was all over on 2 May, 1945. When the final German surrender came in Italy the 16th/5th Lancers were 'harboured round some farm buildings a few hundred yards from the Po'. It was a long, long way from Risalpur in the Punjab where they had been when the war began. The Sherman tanks with their 76mm and 105mm guns had little in common with horse and lance, but the much vaunted 'cavalry spirit' did not seem to have been much affected by the change. The panache and *élan* of the Peninsular War had survived mechanization! It may have been in some way symbolic that Lieutenant-Colonel Dennis Smyly, who was commanding the Regiment at the time of the German surrender, was a first-class horseman who was to play a prominent part later in the development of 'Eventing', now such a popular part of the equestrian scene. Dennis Smyly, a man of great charm, was Colonel of the Regiment from 1959-69.

The German soldier had been as tough an opponent in Italy as he had been in North Africa. From the beginning of the campaign until the end he had been quick to seize the fleeting opportunity in attack, and tenacious in defence. If war is a hard school, the German soldier proved himself to be a tough teacher. The 16th/5th Lancers buried thirty-two of their comrades in Italian soil (four of them officers), and of all the many Battle Honours they have won in their long history, Italy 1944-45 was harder earned than most of the others.

CHAPTER NINE

End of Empire
1946-67

AFTER VE Day in 1945 the Regiment found itself engaged in occupation duties in the British Zone in Austria. After several months it moved to Schleswig-Holstein in Germany for the same purpose. Europe was in a state of turmoil almost impossible to convey to anyone who was not there at the time. Cities had been battered to the ground, road and rail communications disrupted, and hordes of displaced persons of almost every European nationality roamed the countryside and streets. There were Cossacks fleeing from the Red Army, Yugoslav Chetniks trying to escape from Tito's Partisans, Russians, Dutchmen, Poles, Frenchmen and others all on the run from Hitler's concentration camps, and Jews desperately trying to make their way to the USA or Palestine. To add to the chaos were those former members of the SS or Gestapo fleeing from the wrath to come, changing their identities in order to find asylum across the sea in North or South America; some of them are still being sought 50 years after the war has ended. Even Carol Reed's film masterpiece *The Third Man* fails to portray the full horror of those times.

The end of a long war means that a regiment will need to be virtually reconstituted. This was certainly the case with the 16th/5th Lancers. Between the end of the war and the middle of 1949 it had no less than five commanding officers, three of them posted in from outside; one from the 12th Lancers, one from the 5th Royal Inniskilling Dragoon Guards and one from the 10th Hussars. There had been a similar upheaval among the officers and senior NCOs. By June, 1949, there were few officers still serving with the Regiment who had fought in it during the war. Their replacements had come from the Foot Guards, the Infantry (no less than three from the Royal Fusiliers), the Indian Cavalry, and one even from the RAF. There had been a similar turn-around in the Sergeants' Mess. The RSM had come from the 7th Hussars, and other senior NCOs from other cavalry regiments. By far the largest contingent, however, had come from the Royal Gloucestershire Hussars, a Yeomanry regiment which was sent home from Austria to demobilize. Many of their NCOs and Troopers were cross-posted to the Regiment. There had of course been a similar upheaval after the First World War, further complicated then by the amalgamation of the 16th with the 5th Lancers. Inevitably it takes time to re-establish that feeling of corporate identity which is the hallmark of the British Army's regimental system. After both World Wars it took about five years.

The Regiment did not stay long in Schleswig-Holstein before returning home for the first time in five years. It was under orders to go out to the Canal Zone in Egypt as an Armoured Car regiment and staged briefly at Lulworth Camp in Dorset in order to convert to Daimler armoured cars and Humber scout cars. While there, however, a very important event took place. King George VI marked the 21st Birthday of Princess Elizabeth on 21 April, 1947, by conferring on her the Colonelcies-in-Chief of three regiments, the Grenadier Guards, the Argyll & Sutherland Highlanders and the 16th/5th Lancers. This was the beginning of the Regiment's highly prized connection with the Sovereign that will be dealt with at greater length later in this chapter. The Princess paid her first visit to Her Regiment while it was at Lulworth.

There had been great changes in Anglo-Egyptian relations since the Regiment had last served in Egypt in 1926. By the Treaty signed in 1936 it had been agreed that British troops would be removed from metropolitan Egypt to the vicinity of the Suez Canal. The outbreak of war in 1939 had prevented this from happening but it could no longer be delayed once the war was over. There was a great difference between the fleshpots of Cairo and Alexandria and the string of dreary, fly-blown tented camps strung out along the Canal from Port Said to Suez. The camps were surrounded by wide perimeters of barbed wire that even so failed to keep out the expert thieves of the area; even anti-personnel mines failed to deter them. There were few amenities outside Port Said and Ismailia and the attitude of the local population was not particularly friendly. Few who had to endure those hot summers of 1949 and 1950, with the *khamsin* filling the air with sand and blowing the tents to tatters, would choose to repeat them. The Regiment's camp on the Ismailia-Suez road was called Camp 53, but it was known to most of its occupants as Stalag 53.

It was, however, ideal country in which to train a reconnaissance regiment; miles and miles of desert all the way to Cairo. It was a pity that the armoured cars, veterans of many years' campaigning under poor climatic conditions, had very little life left in them. It was not unusual to return from an exercise with the majority of them on tow. It certainly gave the fitters plenty of chance to improve their skills. These exercises provided an excellent opportunity to weld together as a regiment the many new officers and NCOs, as well as the National Servicemen who made up the majority of the Regiment. It was indeed fortunate that at this critical moment in the Regiment's history it acquired a Commanding Officer of great quality, whose breadth of experience and strong personality was well suited to the task he faced. He was Lieutenant-Colonel D. R. B. Kaye, formerly of the 10th Royal Hussars, which regiment he had commanded in the Western Desert, for which he had been awarded the DSO. He succeeded Lieutenant-Colonel T. C. Williamson who had commanded the Regiment for about twelve months.

Douglas Kaye proved to be a remarkably successful Commanding Officer. In the same way as Colonel Cecil Howard put the 16th Lancers back on their feet in the aftermath of the First World War, Kaye was just as successful with the 16th/5th Lancers after the Second World War. He was a big man in every meaning of the word; watching him squeeze into the turret of a Daimler armoured car was an entertainment in itself. He had no time for the pretentious, the inefficient and the idle. He insisted on high standards both on and off parade. Those who failed to match up to them, whatever their rank, were soon on their way. Nor did he in any way kow-tow to higher authority. He could be as outspoken to a brigadier as to an errant troop leader. Needless to say he ruffled the feathers of some, but he was greatly respected by the great majority. He served the 16th/5th Lancers well.*

One of his first tasks was to recommend to the Military Secretary a successor to Colonel Cecil Howard as Colonel of the Regiment. Colonel Howard died on 24 January, 1950, having been Colonel of the Regiment since 1943. His successor was Brigadier P. E. Bowden-Smith who had commanded the Regiment in 1936-37. He was one of the finest horsemen of his generation and played a leading part in establishing the Three Day Event at Badminton, now the premier equitation event of its kind in Britain. He was assisted in this task by several retired 16th/5th Lancers. 'Bogey' Bowden-Smith was for several years Joint-Master of the Pytchley Hunt.

As stated earlier, the honorific appointment of Colonel of a Regiment is a relic of the time in the British Army's history when a regiment was almost the private property of its Colonel, after whom it sometimes took its title. Colonels were then responsible for almost every aspect of their regiment's administration, including the pay of all those listed on the muster roll. Not infrequently those who had died or deserted continued to be paid, the Colonel and the Paymaster pocketing the proceeds. By the end of the last century the appointment had been regularized, becoming the summit of a regimental officer's ambition. It was an honorific one made by the Sovereign on the advice of the Military Secretary, who was in his turn advised by the outgoing Colonel, or by the Regimental Council if there was one. The appointment can best be described as the 'Head of the Regimental Family'. It is customary, although not mandatory, that the holder should have commanded the Regiment. More often than not he will be on the Retired List and whatever his rank, be it Major or

---

*So far as this author is concerned, he learnt more of how to command soldiers from Douglas Kaye than from any other officer under whom he served during thirty-five years in the Army.

Major-General, he will when in uniform wear the insignia of a Colonel and the badges and buttons of his regiment. He receives no pay but is paid a small allowance which barely covers postage. He has no powers of discipline (unless he happens to be still serving) and cannot hold the appointment beyond the age of 70. He chairs all regimental committees, such as the Regimental Association, Benevolent Fund etc, and plays an important part in the selection of potential officers, as well as being the normal channel of communication on regimental matters to the Sovereign and Military Secretary. The appointment of Colonel of a Regiment is peculiar to the British Army, although some foreign armies have adopted it, such as India and Pakistan.

In February, 1950, the 16th/5th Lancers moved from Egypt to Barce, in Cyrenaica in North Africa. There they took over from the 13th/18th Royal Hussars, who in their turn were expected to take over at Fanara in the Canal Zone. A Squadron of the Regiment was to go to Khartoum on a 12-month detached tour and remained behind to hand over the camp at Fanara to the incoming regiment. In the event, however, the

M.V. Dilwara, one of the troopships which ferried soldiers between UK and the Regiment in the Canal Zone and North Africa 1948–53

Inspection of the Regiment by Field-Marshal Sir William Slim at Fanara, Egypt, 1949

13th/18th only staged at Fanara for a few weeks before going out to Malaya where the Emergency was at its height. Had the decision to send a second armoured car regiment to Malaya been taken before the Regiment had left for Barce, it would have been the Regiment which was chosen to go. As it was, A Squadron remained at Fanara for the rest of the year until relieved early in 1951 by the Royals. Its operational task in the event of hostilities with Egypt was to seize the polo ground at Gezira in the heart of Cairo as a dropping zone for British paratroops. Since there was a complete Egyptian armoured division equipped with Centurion tanks barring the way to Cairo, against which the 2-pounder guns of Daimler armoured cars would have had much the same effect as a pea-shooter, it was as well that A Squadron was never put to the test.

Barce was an improvement on the Canal Zone, although the former Italian barracks were in an advanced state of dilapidation. The accommodation for the families was, however, a great improvement on Egypt. The climate was more temperate and Barce itself for much of the year was green, although the great Libyan desert was at its very doorstep. It provided excellent training for a reconnaissance regiment, although great care had to be taken throughout the terrain.

 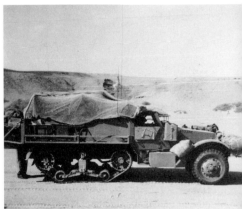

The Regiment was equipped with Daimler Armoured Cars and Scout Cars in the Canal Zone 1948–50

The Half Track – an ideal vehicle for the dese – in service with the Regiment 1948–53

This was where the Eighth Army and Rommel's Afrika Korps had fought their great battles and the countryside was littered with the hulks of tanks and buried anti-tank and anti-personnel mines. One squadron was at Derna on the sea coast which provided opportunities for swimming. Barce was in fact a pleasant garrison but the constant state of tension in the Middle East kept the Regiment on its toes. In October, 1951, it was moved at very short notice fully 'bombed up' to the Cyrenaica-Egypt border ('The Wire'), ready to move into Egypt should hostilities break out. Anglo-Egyptian relations were growing worse and worse and there was a good deal of sabre-rattling by both sides. Fortunately the Regiment was not called upon to move, but on 26 January, 1952, the Cairo mob rose and burnt and sacked Shepheard's Hotel and the Turf Club in Cairo, killing several British nationals. This was followed in July by the overthrow of King Farouk and the rise to power of Colonel Gamal Abdel Nasser. It was the end of the era of British power in the Middle East, although very few realized it at the time.

The 16th/5th Lancers were not, however, to be further involved. In December, 1951, Lieutenant-Colonel Kaye handed over command to Lieutenant-Colonel J. R. Cleghorn who was

Lieutenant-Colonel Cleghorn and Lieutenant Anderson with the pack of fox
hounds which the Regiment had at Barce

Exercising in the djebel from Barce in springtime

no sooner in the saddle than he was informed that the Regiment
was to move to Zavia, some 27 miles west of Tripoli, where it
was to convert from its reconnaissance role to that of anti-tank
defence. This involved giving up its armoured cars for Comet
tanks and put a premium on the skills of tank gunnery. Part of
the Regiment moved to Zavia by road, a distance of roughly 700
miles, and another part by sea in LSTs. The forty-odd families

On arrival in Zavia (near Tripoli) the Regiment re-equipped with Comet tanks, 1952

were moved by air. The barracks at Zavia were a great improvement on Barce, although Zavia itself was not as attractive as Barce; but in the event it did not matter much. Hardly had the Regiment settled in and started to train for its new role than Colonel Cleghorn was informed that the Regiment would move to Germany early in 1953, staging briefly in England en route. It would, moreover, change role yet again, this time to operate as an armoured regiment equipped with the Main Battle Tank, the Centurion, a great advance on the Comet.

After 1947, when India and Pakistan became independent, it might have been thought that the calls on the British Army for service overseas would be significantly diminished. The average strength of the British Army in India over the years had been between sixty and seventy-five thousand, roughly one British soldier for every 3,800 Indian ones, so the War Office looked forward to a considerable reduction in overseas commitments. However, it did not work out that way, partly because there was no longer a British-controlled Indian Army to help shoulder the burden in places like Aden and Malaya. There was also the requirement to maintain an army of occupation in Germany that later became the British contribution to NATO, and very nearly equalled in numbers the former British garrison in India. In fact the demand for soldiers was never-ending, but for a variety of reasons the strength of the Army continued to decrease, particularly after the ending of National Service in 1961. Even before then the opportunities in civilian life were growing more attractive, which inevitably affected regular recruitment, as did in a fashion the effects of National Service. The frequent comings and goings in what was a conscript army had an unsettling effect on the regulars; a tank crew was no sooner trained and exercised than it would be necessary to begin all over again with a new crew. 'Turbulence', or 'Over-Stretch', was to be the soldier's lot for many years to come.

The 16th/5th Lancers, for example, had had no less than three different roles, which meant training in three different kinds of equipment, in less than three years. This particularly affected the long-service NCOs, the cement that held together an army of National Servicemen, and it affected not only them but also their families, who were subjected to frequent moves. The requirement for officers of captain's rank and above to fill appointments on the Staff or at Schools of Instruction etc. made it difficult, if not impossible, for a Commanding Officer to plan ahead to fill such important appointments as Squadron Leader. Although too long a stay in one garrison has its disadvantages, too frequent moves can be equally bad for morale and efficiency. Since 1945 the Regiment had been in Austria, Germany, UK, Egypt, Cyrenaica, Libya, and was now going back to Germany again.

When it arrived in West Germany in July, 1953, after staging for a few months at Quorn Camp in Leicestershire, the German people were still recovering from the traumatic effects of defeat. BAOR was still very much an army of occupation and the victors claimed many privileges over the conquered. It was still possible to hold large-scale manoeuvres across the countryside without arousing unfavourable comment among the civilian population, and British troops enjoyed many valuable privileges such as cheap petrol. The 'German Miracle'

On exercise in BAOR

had yet to take off and there was no doubt in most people's minds that the Soviet Union was *the* potential enemy. This gave service in BAOR a sense of purpose that was to diminish with the years, but it was very much there in 1953.

Athlone Barracks in Sennelager was to be the Regiment's home for the next four years. Sennelager had been one of the *Wehrmacht's* principal training areas, and it was there that Field-Marshal Guderian first tried out the armoured/infantry tactics that were to prove so successful in France in 1940. The Regiment had excellent tank-training facilities on its very doorstep. It formed part of 33rd Armoured Brigade in the 11th Armoured Division, whose commander, Major-General H. E. Pyman, was one of the leading exponents of armoured warfare in the British Army. Training was hard, and almost continuous. The Centurion tank was a vast improvement on the Comet, its drawback being its slow speed and poor operational range. It was in the process of being up-gunned with the 105 mm gun which has proved itself to be one of the best tank guns in the world.

Most of the tank-equipped regiments in the Royal Armoured Corps, both Cavalry and Royal Tank Regiment, were in BAOR, and the competition was intense. This applied not only to military training but involved sport of every kind, including skiing, sailing and polo. At the big events such as Horse Shows, Race Meetings etc, regiments vied with each other when it came to the entertainment of guests. In this respect the Regiment was fortunate, having in Anthony Bullivant, first as Second-in-Command and then as Commanding Officer, an impresario of considerable gifts. He was ably seconded by the Officers' Mess Sergeant, Stanley Russell, a Head Chef of great talent and a *Maître d'Hôtel* of almost classic proportions, in girth just as much as in other ways. He was probably the best known Mess Sergeant in BAOR. Every General knew him!

It was in 1954 that the Regiment acquired a new title which had been conferred on it by the Queen. When King George VI died in 1952 and Princess Elizabeth ascended the throne, it was to be expected that, as was customary, all previous appointments held by Her would be deemed to lapse, including of course Her Colonelcy-in-Chief of the Regiment. To the Regiment's great delight, however, the Queen indicated Her intention of remaining as Colonel-in-Chief of the 16th/5th Lancers, and moreover conferred on it the new title of the *16th/5th The Queen's Royal Lancers*. This first appeared in the Army List in August, 1954. In that same year a new collar badge, combining the badges of the 16th and 5th Lancers was taken into use. It is known as *The Queen's Badge*.

Lieutenant-Colonel Cleghorn handed over to Lieutenant-Colonel Anthony Bullivant at the end of 1954. During his three years in command the Regiment had had three changes of role and no less than four moves. Bullivant's succession was significant in more ways than one. He was the last officer on the regimental list to have served with the Regiment while it was still horsed; he joined in Risalpur in India (now Pakistan) in 1936. He had served with the Regiment throughout the North African and Italian campaigns, apart from short spells away on the staff, during which he had held every officer's appointment in the Regiment other than that of Commanding Officer. He was to complete this record in 1969 when he was appointed Colonel of the Regiment, an appointment he held until 1975.

It was during Bullivant's period in command that the 11th Armoured Division, with its divisional sign of a charging bull, was demobilized. The 16th/5th Lancers then became the armoured regiment supporting 4th Guards Brigade, most of the other units being located in the vicinity of Dusseldorf. The Regiment, however, was halfway to the East German border at Sennelager. At this time there was considerable debate as to how best to deploy BAOR to counter a Soviet invasion of West Germany. One school of thought favoured a fall-back to the Rhine as the best obstacle on which to fight the Russians. The other argued for fighting as far forward as possible. This was the policy finally adopted, which is hardly surprising once West Germany had joined NATO in 1955. It was unthinkable for any German to consider abandoning the greater part of the Federal Republic in order to halt the Russians on the Rhine. It also made more sense for the 16th/5th Lancers. It was difficult to imagine how they would manage to get back to join up with 4th Guards Brigade along roads clogged by refugees and under constant attack from the air. It would be 1940 all over again!

The Regiment returned home in 1957 with yet another change in role. It was to form part of the RAC Training Brigade based at Catterick Camp in Yorkshire. During, and for some years after the war, recruits for the various regiments of the Royal Armoured Corps were trained by units specially formed for the purpose. But as the size of the British Army was drastically reduced from 1957 onwards, it was decided to avoid disbanding or amalgamating more regular regiments by making each regiment in turn serve in the Training Brigade (of four regiments), with the task of training recruits for certain specific R.A.C. regiments. It can hardly be described as a very glamorous role but it was an important one, and it certainly helped to

Regimental Sergeant Major Jock Marshall MBE DCM instructing National Service recruits at Catterick while the Regiment filled the role of Royal Armoured Corps Training Regiment 1957-59

keep regiments in being. It did, however, entail a complete reorganization within a regiment, each sabre squadron being made responsible for teaching one of the three basic Armoured Corps skills, viz: driving and maintenance of tanks; gunnery; and radio operating. Additionally there had to be a Wing in which selected recruits could undergo training as potential officers before they attended the War Office Selection Board at Barton Stacey for final selection for officer training.

The Regiment's arrival in Catterick coincided with a major political decision made by the Conservative Government of Harold Macmillan. This was to phase out National Service by 1961 and return to the pre-1939 arrangement of an all-regular army. Along with this change in policy there was to be a significant reduction in the size of the regular army which, in the case of the Royal Armoured Corps, meant the loss either by disbandment or amalgamation of eight units. Fortunately the Regiment was not to be one of them. It did mean, however, that the problem of recruiting regular soldiers had to be given the highest priority, and this was a subject to which a great deal of attention was given during the Regiment's stay in Catterick, fortunately to good effect. On more than one occasion between 1961 and 1990, the 16th/5th Lancers has been the best recruited regiment in the Royal Armoured Corps.

Much of this success can be attributed to the close affiliation the Regiment has managed to achieve with the County of Staffordshire. Cavalry regiments in the past had no particular connection with specific parts of the United Kingdom, apart of course from the Irish regiments and the Royal Scots Greys; otherwise they drew recruits from all over the country. In 1947, however, it was decided to affiliate regular cavalry regiments with the Yeomanry regiments of the Territorial Army. The 16th/5th Lancers were affiliated with the Staffordshire Yeomanry (Queen's Own Royal Regiment), with headquarters at Burton-on-Trent in Staffordshire. For nearly twenty years, until the Staffordshire Yeomanry disappeared as a result of the reorganization of the Territorial Army in 1966, the 16th/5th Lancers provided the Yeomanry with its permanent staff. This consisted of a major, a captain (as adjutant), the quartermaster, RSM, and several senior NCOs and troopers. Although the Regular Army has been accused of failing to treat the Territorial Army with the respect that it merits, this was certainly not true of the relationship built up between the 16th/5th Lancers and the Staffordshire Yeomanry. No less than five future commanding officers served with the Yeomanry, either as majors or captains, and when the Staffordshire Yeomanry was resurrected in 1971 as a squadron of the Queen's Own Mercian Yeomanry, the Regiment was delighted to be called upon to provide the squadron with its Permanent Staff Instructor (PSI).

The Yeomanry has its origins far back in British history. It used to consist chiefly of the 'Gentlemen' of the county who brought with them their tenants. The great landowners treated their regiments as their own. The Earl of Dudley, for example, who took the Staffordshire Yeomanry to war in 1914, was reputed to spend £3,000 or £4,000 a year on maintaining his regiment. Disapproving of the Government-issue sword, he armed every officer and trooper with a sword of his own invention. Strict instructions were issued that a wounded or sick soldier would be relieved of his 'Dudley' sword before being evacuated to hospital. In both world wars the Staffordshire Yeomanry distinguished itself and its Battle Honours include both El Alamein and the Crossing of the Rhine. When it was decided to take a leaf out of the Infantry's book and link Cavalry and RTR regiments with counties, naturally the Regiment was affiliated with Staffordshire, establishing a Home Headquarters first at Burton-on-Trent, and since 1968 at Stafford, where the Regimental Museum has also been established. Over the years the association with Staffordshire has been strengthened until today a high proportion of soldiers come either from Staffordshire or the adjacent county of West Midlands. It is indeed a sign of the times that the annual dinner of the Regimental Association, which for more than 100 years has been held in London, will from 1990 onwards be held in Birmingham, from which region so many of the soldiers are recruited.

Catterick was not a very popular garrison, due as much to the appalling state of the barracks as to its very severe winters. There was, however, a great deal of sport, such as hunting and shooting, to be had close at hand and the local people were very friendly. Organized sport was more difficult because the National Servicemen liked to get home for weekends, whenever possible. This reduced the scope for organized games. However, some very talented sportsmen passed through the training machine which enabled the Regiment to build up an outstanding football team which won the Army Cup for football in April, 1959. This was the first time it had ever been won by a cavalry regiment.

There were two events connected with the Queen that in some fashion made all the difference to what might otherwise have been a rather humdrum period in the Regiment's history. The first was on 10 July, 1957, when the Queen paid her first visit to the Regiment since Lulworth ten years previously. She was able to visit all the Training Wings and took the liveliest

H.M. The Queen presents the Guidon in the garden of Buckingham Palace, March 1959. Painting by Sir William Dring RA

The Queen dancing with Brigadier Bowden-Smith, Colonel of the Regiment, following the Guidon Parade. *The London News Agency Photos Ltd*

interest in everything she saw. Then on 19 March, 1959, the Queen presented the Regiment with its Guidon on the back lawn of Buckingham Palace, some 200 officers and men being present on parade. That same night the Queen attended a Dinner and Ball given by the officers at the Hyde Park Hotel to mark the bicentenary of the raising of the 16th Light Dragoons. The Duke and Duchess of Gloucester were also present. It was a memorable occasion.

Guidons, it will be remembered, are swallow-tailed flags carried by Light Cavalry; the Household and Heavy Cavalry regiments carry Standards, which are square. Infantry regiments carry Colours. The Battle Honours of a regiment are inscribed on these and they are consecrated before presentation. Light Dragoon regiments which were converted to Lancers in 1816 no longer carried a Guidon, their Battle Honours being inscribed on the Drum banners that were draped round the two mounted Kettle Drums. However, in 1958 it was ordered that

Lancer regiments, like Hussars, would carry a Guidon. There was some difficulty when it came to deciding the centre-badge of the Regiment's Guidon. This is usually the regimental badge, but for some heraldic reason this could not be done in the Regiment's case. Bearing in mind that the first Guidon of the 16th Light Dragoons had the cypher of Queen Charlotte as the centre badge, it was decided that the centre badge of the new Guidon should be the cypher of Her Majesty Queen Elizabeth II – 'ER II'. Her Majesty was pleased to approve of this.

Regulations lay down that Guidons, Standards and Colours are to be replaced every 25 years. Therefore on 15 July, 1983, the Guidon presented in 1959 was laid up and a new Guidon was presented by the Queen at Tidworth, where the Regiment was then stationed. It was a remarkable parade by any standard, combining one squadron dismounted carrying lances, and two squadrons in Scorpion light tanks. The drill was impeccable and after presenting the Guidon the Queen took lunch with the entire Regiment, their families and their guests, amounting to 1,900 people in all. Afterwards the Queen toured the Regiment and took tea in the WO's and Sergeants' Mess before leaving. It must have been one of the most remarkable, and successful, events in the Regiment's history.

There cannot be the slightest doubt that all ranks of the 16th/5th Lancers particularly pride themselves on their close connection with the Queen, their Colonel-in-Chief. As anyone who has had the honour of an audience with the Queen can endorse, her interest in the Regiment, and her knowledge of its activities, is remarkable; the Queen is, after all, Colonel-in-Chief of many regiments besides the 16th/5th Lancers but she has always made a point of meeting the Regiment and its families whether on a formal occasion such as at Tidworth, or informally at a tea party in St James's Palace. Both the Colonel of the Regiment and the Commanding Officer report annually on the state of the Regiment to the Queen, and the incoming and outgoing Colonels of the Regiment are always received in audience by the Queen. The Regiment feels very much that they are 'The Queen's Royal Lancers'.

H.M. The Queen arriving at the Ball following the Guidon Parade wearing the brooch which was presented to Her Royal Highness The Princess Elizabeth by the Regiment on her appointment as Colonel-in-Chief in 1947. *Daily Express Photograph*

After two years spent training recruits at Catterick, the Regiment's return to BAOR in April, 1959, as the armoured regiment in 12th Infantry Brigade Group at Osnabruck proved to be a testing experience. The Regiment had virtually to be rebuilt from scratch, but it did have several advantages. Firstly, all those NCOs and troopers employed as instructors had become very skilled in their RAC trades. Secondly, several senior NCOs from other regiments who had been attached as instructors elected to remain with the Regiment. Thirdly, some excellent National Servicemen and regular soldiers had chosen to join the Regiment. With this to build on the Commanding Officer, Lieutenant-Colonel R. A. Simpson, who had been Second-in-Command at Catterick, was able to produce a well-trained regiment surprisingly quickly. Lieutenant-Colonel Simpson had been badly wounded when serving with the 16th/5th Lancers in North Africa, and was again severely injured when run over by a car when serving as Brigade Major of 33rd Armoured Brigade.

1960 and 1961 were years of consolidation after reorganizing from the training role. There was also the imminent ending of National Service and the need to recruit regular soldiers to take their place. Most difficult of all, perhaps, was the constant comings and goings of officers to or from the staff. By the end of 1961 there were only seven officers (including the two Quartermasters) who were holding the same appointments as they had been 12 months previously. The Regiment had no less than three officers at the Staff College, a great change from the past when in both the 16th and the 5th Lancers attendance at the Staff College was certainly not encouraged.*

The 'Sixties were in Britain, and elsewhere, the 'Years of Change', when long-observed conventions were abandoned, and when young men and young women were no longer content to do as they were told. This change of attitudes in British society inevitably had its effects in the British Army which is only a mirror of the nation from which it recruits its ranks. More than a quarter of the Regiment were now married and could come and go as they pleased when off duty. It therefore made no sense to impose restrictions on their

* Both Generals Gough and Beddington of the 16th Lancers had attended the Staff College but they were exceptions to the general rule of their times.

comrades who lived in barracks and such time-honoured impositions as 'marking-out' were abolished. NAAFI canteens were allowed to sell spirits and soldiers were permitted within limits to choose the colour scheme for their barrack rooms. Most noticeable of all, perhaps, was the enormous increase in privately-owned motor cars. With all this change happening around him, the British soldier was changing too. He was much more aware of what was happening in the world, and better educated than in the past. The terrorist threat had yet to manifest itself and to that extent the soldier was much freer in his off-duty hours than can be permitted today. There was every form of sport available to him, as well as sailing, skiing and horses. And all around him was evidence of the 'German Miracle' as the people of the Federal Republic rose like a phoenix from the ashes to build the most prosperous economy in Europe. There was even a new German Army, the *Bundeswehr*, beginning to take shape, soon to become, after the Americans, the most powerful army in NATO. As further evidence of the importance of that international alliance, for 18 months the 16th/5th Lancers had an American officer attached to it, Major Raymond Beaty of the US Armoured Corps. He commanded A Squadron and made a great success of it. In return Major Richardson was attached to the 7th US Cavalry, 'Custer's Own', with which regiment, as it happened, the 16th/5th Lancers had something of a connection.

When, in 1866, soon after the ending of the Civil War, the 7th US Cavalry were first raised for service on the Indian Frontier, the officer assigned to command it was Lieutenant-Colonel George Armstrong Custer who had reached the acting rank of Major-General at the age of twenty-five in the Union forces. The soldiers were a tough, colourful and experienced body of men, most of them having served during the Civil War. Among them were many Irish immigrants, some of whom had deserted from the British Army, some undoubtedly from the 5th Lancers. 'Garry Owen' was then a popular song among the Irish and one day Custer heard some of his soldiers singing or whistling the song. He liked it so much that he adopted it as the regimental march and the 7th Cavalry's nickname today is the 'Garry Owens'. What is more, there is a town in Montana, not far from the Little Big Horn, where Custer was killed in 1876, which is called Garryowen.

Just prior to the Regiment's move to BAOR the Colonelcy of the Regiment had changed, Brigadier Bowden-Smith handing over to Colonel Dennis Smyly who had commanded the 16th/5th Lancers during the final battle in Italy in 1945. A notable departure from the Regimental List in 1961 was that of Colonel Frank Watson, a 'Scarlet Lancer' in the Robertson tradition. He had joined as a trooper in 1931, rising rapidly through the ranks to become a Sergeant-Major Riding Instructor by 1937 when aged only twenty-four. He was commissioned in 1942, winning the Military Cross in North Africa, and later the Distinguished Service Order in command of B Squadron in Italy. He attended the first post-war course at the Staff College in 1945-46, holding thereafter a wide variety of staff appointments, culminating as Commandant of the RAC Gunnery School at Lulworth. Frank Watson, like Robertson before him, was a natural soldier, and had he been older when war broke out, might well have risen to higher rank during the fighting. There were fewer opportunities for him to do so under peacetime conditions. Frank Watson died in 1991.

Probably the Regiment's greatest achievement in 1962 was that its recruiting campaign had been so successful that the War Office saw fit to impose a limit on the numbers it could recruit. It was now at full strength and all the time and effort expended on building close relations with Staffordshire had produced dividends. The achievement is the more remarkable on account of the fact that jobs were not in short supply in Staffordshire and it was not a case of joining the army for want of employment. The best Recruiting Sergeant is the contented soldier, and many of the men who joined the Regiment at that time did so as a result of advice from friends or relatives already serving in the Regiment.

The Regiment had now been almost four years in BAOR and there were rumours of a move farther afield. Lieutenant-Colonel Bull, who had taken over from Lieutenant-Colonel Simpson as Commanding Officer, was told officially that this would happen in December, 1963, but not as a complete regiment. Regimental Headquarters, Headquarters Squadron, and one sabre squadron would go to Aden; one sabre squadron to Bahrein in the Gulf, and the third sabre squadron to Hong Kong. Only the Hong Kong squadron would be permitted to take families. The tour would be for twelve months. This interesting period in the Regiment's history, which included participation in almost the last of Britain's colonial wars, will be the subject of a later chapter.

1963 was the year General Sir Hubert Gough died. He was ninety-three and certainly the most distinguished 'Scarlet Lancer' of his generation. When he was promoted in May, 1916, to command an Army in France he was not yet forty-six. It was his misfortune to be made the scapegoat for his Fifth Army's failure to halt the German offensive in March, 1918, but he was handsomely exonerated when the Official History was published nearly 20 years later. Another World War I veteran who died in the same year was Lieutenant-Colonel H. A. Cape, the last Commanding Officer of the 5th Royal Irish Lancers. He had visited the Regiment when it was at Catterick and although in his eighties had only just given up his job as Inspector of Steeplechase Courses. He impressed everyone he met by his vigour.

When the Regiment reassembled in Tidworth in February, 1965, after being split up in squadron detachments across the globe, it must have come as a great relief to Lieutenant-Colonel Holland who had succeeded Lieutenant-Colonel Bull early in 1964. He had spent much of that year flying to visit his scattered command. At Tidworth the Regiment was equipped with Centurion tanks, B Squadron being detached at the School of Infantry at Warminster as the demonstration squadron. This was a very demanding task, their tanks averaging 1500 miles in the year, nine of their sixteen having exhausted their normal mileage-life by the end of the detachment. A and C Squadrons had their excitements too, A being flown to Libya on an exercise, and B to Canada for the same purpose with Lord Strathcona's Horse. The End of Empire has certainly not meant that the modern British soldier no longer 'Sees the World'!

Soon after the Regiment's arrival in Tidworth it acquired a yacht. It was a 30-foot glass fibre boat which slept four and just rated as a Class III ocean racer. It was named the *Queen Charlotte* and moored at Marchwood, off Southampton. It proved to be a most valuable acquisition, providing many officers and soldiers with experience in seamanship over the years until it was finally laid up some ten years later. There may not at first sight appear to be much in common between deep sea sailing and crewing a tank but in both cases the need is uppermost for each member of the crew to 'muck in'; what is more the sea is a stern teacher. By the end of the 1965 sailing season no less than three officers and three NCOs and troopers had been graded as Skippers (Class I).

Ever since the end of the war the Regiment had relied on the permanent staff of the Staffordshire Yeomanry to provide it with a home base while the Regiment was overseas. From 1952 onwards this task was most ably undertaken by Major Warner who had joined the Regiment in Austria as Quartermaster from the Royal Gloucestershire Hussars. In 1952 he handed over as Quartermaster to Captain Bowman and went to the Staffordshire Yeomanry as Quartermaster. A Staffordshire man himself, Warner was extremely successful in establishing the Regiment's links with the county and played a leading part in setting up the Regiment's Home Headquarters alongside the Yeomanry's, first at Burton-on-Trent, and from 1968 at Stafford. Much of the Regiment's recruiting success was due to him. He took a prominent part in local government and was about to become Mayor of Burton-on-Trent when, unfortunately, he died in 1977. He had handed over at Home Headquarters to Lieutenant-Colonel R. A. Bowman in 1971.

It would be impossible to over-estimate the importance of a Home Headquarters to a cavalry regiment that hitherto had lacked the means of keeping in touch with important matters at home while it was serving overseas. This became doubly important once cavalry regiments were linked with counties and when success in recruiting largely depended on the strength of such links. When this was recognized by the Ministry of Defence and Home Headquarters were established for each cavalry regiment, with a retired officer acting as regimental secretary, it became possible for cavalry regiments to establish the same close links with their counties as infantry regiments had previously been able to do. As World War II veterans grow older, the number of welfare cases increase, and care for the Regimental Benevolent Association is one of the regimental secretary's principal responsibilities. He is also responsible for overseeing the Regimental Museum now set up at Kitchener House at Stafford, and for keeping in touch with the county authorities such as the Lord-Lieutenant's office, the Constabulary, the Army Cadet Force, the local media and the Territorial Army.

The Territorial Army was reorganized in 1966 which gave it the role of reinforcing BAOR and bringing it up to war establishment in the event of war. This resulted in wholesale reductions which included the Staffordshire Yeomanry, which disappeared from the Army List. One of the new Yeomanry regiments formed was the Queen's Own Mercian Yeomanry in 1971, its B Squadron being based in Stafford and forming a link with the former Staffordshire Yeomanry. Although the reorganization of the Territorial Army was bitterly opposed at the time, it cannot be denied that in its new guise it is a great deal more efficient and better value for money than its predecessor was. If events in Europe, including the reunification of West and East Germany, should lead to a drastic reduction in BAOR, there can be little doubt that the Territorial Army will be reorganized yet again.

The British Army was not unaffected by the remarkable sociological change that took place in Britain during the 1960s. 'Full Employment' was the slogan of both Conservative and Labour Governments which had its inevitable effect on recruiting. Furthermore, women increasingly joined the labour force, whether married or unmarried. When wives found themselves in well-paid jobs, adding considerably to their families' incomes, it is not surprising that they should be reluctant to leave them when their husbands were posted abroad. Moreover, home ownership became another political slogan which meant mortgages that had to be serviced. The trained and disciplined serviceman discovered that the skills he had acquired in the army were also much in demand in industry, and at a much higher rate of pay. The result was a constant drift to civilian life, often by young officers and senior NCOs who were very difficult to replace. The nomadic nature of the soldier's life that had once been one of the attractions of soldiering now became something more than a drawback where wives were concerned. It could mean literally the halving of a family's income when moving to Germany where employment for British women was much more difficult to find. Separation for either shorter or longer periods that once was accepted as the soldier's lot was increasingly unpopular. These were problems that every Commanding Officer had to face whenever a regiment was ordered overseas, and they have not lessened with the years.

However, overseas service, apart from West Germany, was fast coming to an end, although there still remained a few hangovers of empire, such as the Falklands and Belize, that required attention. When the last British soldier left Aden on 29 November, 1967, the long years of empire seemed to have finally ended. The great cavalry stations of the past − Cairo, Risalpur, Sialkot, Ambala and Secunderabad − have become no more than memories, and it is doubtful if there are any 16th/5th Lancers serving today who could find them on the map!

CHAPTER TEN

The Professionals
1968-1990

THERE were, not unnaturally, many soldiers, both serving and retired, who considered that the end of overseas service (apart from Germany) would have a drastic effect on recruiting. That this did not turn out to be the case − the 16th/5th Lancers were over-recruited in the 'Seventies − was partly due to the effort made by regiments to recruit young soldiers, and partly to the publicity that resulted from the Army's involvement in Northern Ireland from 1969 onwards. The British public saw a great deal more of its soldiers on television than it had ever done when they were far away overseas, and the effect overall was a favourable one. Moreover, the recruiting slogan that was coined at this time − 'Join the Professionals' − laid emphasis on the fact that soldiering was a profession like many others, and that the British Army was very good at it.

When the Regiment left Tidworth for Fallingbostel in January, 1968, Lieutenant-Colonel Brooke was the Commanding Officer. He would be the last in a succession of four Commanding Officers − Simpson, Bull, Holland and Brooke − who had fought in tanks in the war. Holland and Brooke were awarded the MC as troop leaders in Italy. Brooke was the last of the line; all Commanding Officers after him were of the post-war generation.

Fallingbostel is on the edge of the North German plain, equidistant from Bremen and Hamburg. It is also close to the Hohne tank gunnery ranges and the Soltau training area. This makes for good training facilities, but Fallingbostel was not a popular garrison. The barracks were badly heated and furnished. There was a constant demand for barrack fatigues which are the British soldier's special bane. To make matters worse there was a shortage of accommodation for families, which meant the majority had to be housed in Hanover, two hours' travel away by road.

The Regiment was one of the two armoured regiments in 11th Infantry Brigade; the Royal Scots Greys was the other. There were also two Infantry battalions in the brigade. The Chieftain tank was just coming into service to replace the Centurion and the 16th/5th Lancers was one of the first regiments to be equipped with it. It weighed 50 tons, had a road speed of 30 mph, and a range of 280 miles. The gun was 120 mm with a 7.62 mm machine gun mounted coaxially. There was a second 7.62 mm machine gun mounted in the commander's turret. The tank was fitted with a full range of night-visual equipment, including an infra-red searchlight mounted on the left side of the turret. The Chieftain was in every way a much more sophisticated tank than the Centurion. The Regiment also acquired an Air Troop, six Sioux helicopters, for reconnaissance purposes.

One of the most marked changes in the post-1967 British Army has been the more sensible attitude towards Dress. Whereas there was a time when the British cavalryman was so tightly trussed that he could not remount his horse once he had dismounted, voluntarily or otherwise, the introduction of Combat Dress, as well as the sensible pullover and trousers for everyday use in barracks, has been a vast improvement on the past. Although the colour of a tunic, or the title of a regiment, can arouse more controversy among the generals of the Army Board than almost anything else, the British Army today is as sensibly dressed as it has ever been in its history. Admittedly the reintroduction of mess kit, which in the Regiment's case is particularly expensive, has been regarded by some as a retrograde step, just as the retention of the Blue Number 1 Dress for ceremonial occasions has been fiercely criticized by some who would prefer either khaki or the dark green favoured by the Canadian Forces. But compared with George IV's flights of fancy in the early years of the last century, uniform today is both practical and much less costly. When officers wear Service Dress (khaki) in barracks today they no longer wear the regimental pattern Sam Browne belt, unless on parade.

Soldiers still, however, like to be well turned out for the special occasion. An interesting example of this is the wearing of mess kit in the WOs' & Sergeants' Mess for special Dinner Nights, which is now being copied by the Corporals. The mess kit, a less expensive copy of the officers', has still to be bought by the individual on a voluntary basis, and this despite the fact that in Britain as a whole informality in dress is a fact of life. No longer are officers expected to wear dark suits and stiff collars when in London, and of course always black shoes. Bowler hats are as rare as roses in midwinter, except for the Cavalry Memorial Parade in early May. One distinguished cavalry general who boasted that he had never boarded a London bus in his life, nor been seen carrying a parcel in public, would be regarded nowadays as wildly eccentric. Mufti, or plain clothes, or 'civvies,' or however one would choose to describe it, is as likely to be jeans, a turtle-necked sweater and a blazer as anything more formal.

In March, 1969, Colonel Bullivant succeeded Colonel Smyly as Colonel of the Regiment; and in the same month Lieutenant-Colonel Pownall succeeded Lieutenant-Colonel Brooke as Commanding Officer. It was under his command that the Regiment, representing BAOR, won the Canadian Army Trophy for Tank Gunnery at Hohne in 1970. Three armies took part in the competition - Canadian, West German and British — and it became a duel between the British Chieftain and the West German Leopard. The Canadian entrant, Lord Strathcona's Horse, was equipped with Centurions.

One of the subalterns who joined the Regiment while at Fallingbostel was HRH Prince Alexander of Yugoslavia, son of ex-King Peter, whom he has since succeeded. Prince Alexander was a particularly good skier, almost of international standard, and during his time with the Regiment the Regimental Ski Team made its mark in the annual Army competitions. Also while in Fallingbostel great use was made of the *Queen Charlotte* which

Lieutenant-Colonel John Pownall with General Sir Desmond Fitzpatrick after the presentation of the Canadian Army Trophy for tank gunnery in 1970, competed for by the armies of NATO

was berthed at Kiel, comparatively close to Fallingbostel. Many officers and soldiers had their first taste of ocean cruising in the Baltic.

One of the advantages of the British Army's regimental system is the family spirit which it engenders. The 16th/5th Lancers have long prided themselves on this family spirit; at one time in the early 1960s the sons of no less than five former commanding officers were serving with the Regiment, as well as the grandsons of Generals Gough and Geoffrey Brooke, and Colonel Cape of the 5th Lancers; the same father-and-son connection was to be found in the WOs' & Sergeants' Mess. But families have their good and bad times, as do regiments, and the years 1964-70 witnessed the departure of many familiar faces who had joined the Regiment soon after the end of World War II. They left the Army to make their fortunes in civilian life. And there were some who died, such as Brigadier Bowden-Smith in 1964, Major-General Geoffrey Brooke in 1966, Brigadier-General Sir Edward Beddington in 1966, and Lieutenant-Colonel Trevor Horn in the same year. Beddington in his day was considered by some to be the 'cleverest man in the British Army', and his unpublished account of his life (in the regimental archives) is a document of great historical importance, particularly in connection with the dismissal of General Gough in 1918; Beddington was on Gough's staff and a great admirer of his. Bowden-Smith, Brooke and Horn were outstanding horsemen of Olympic class.

The Regiment changed both its role and Commanding Officer at the end of 1971, Lieutenant-Colonel Dennis succeeding Lieutenant-Colonel Pownall. The change of role cuased by the posting 16th/5th Lancers' to Northern Ireland to take over Lisanelly Camp in Omagh (County Tyrone) from the 17th/21st Lancers. This meant changing from an armoured regiment to an armoured reconnaissance regiment, exchanging tracked vehicles for wheeled. The Regiment was now to be equipped with the Saladin, a six-wheeled armoured car with a road speed of 45 mph and a range of 400 km. It was armed with a 76 mm gun and a .3 machine gun mounted coaxially. It was a vast improvement on the Daimler with which the Regiment had been equipped when last an armoured car regiment, as indeed was the Ferret Scout car on the former Humber model.

On 7 October, 1971, shortly after the return from West Germany and prior to the move to Northern Ireland, the Queen attended a regimental tea party in St James's Palace. All the serving members of the Regiment, together with their wives, or girl friends, met the Queen quite informally as she moved from room to room, displaying a remarkable interest in everyone she met. For the very few older warriors present (previous Colonels of the Regiment and Commanding Officers had also been invited) it presented an interesting contrast between the soldier of today and the soldiers of, say, fifty years ago. Then the majority of soldiers would stand tongue-tied when meeting senior officers, whereas the modern 16th/5th Lancer engaged the Queen in vigorous conversation, which Her Majesty obviously enjoyed. It was an extremely happy occasion.

Northern Ireland was to present the Regiment with an entirely new situation, and it remains today, as it did then, the British Army's Number One Problem. As such it merits a separate chapter which will follow later, as will a chapter on the Regiment's involvement in Cyprus. It will therefore be necessary to skip nearly five years and take up this narrative again in 1975 when the Regiment was back in BAOR, still as an armoured reconnaissance regiment, but this time at Wolfenbüttel on the border with the German Democratic Republic (GDR), with the 'Iron Curtain' that stretched from the Baltic to the frontier with Czechoslovakia plainly visible. No one could possibly have imagined then that it too would so soon be passing into history.

Wolfenbüttel's proximity to the high barbed wire fences strewn with anti-personnel mines that divided the two Germanies gave service there a sense of purpose that was not always present in other BAOR garrisons. The Regiment's regular patrolling of the border was carried out under the gaze of the East German border guards from their tall watch-towers. It may also have accounted for the friendliness of the local inhabitants who were only too well aware of the brutality of the Communist régime beyond the frontier fence. Northampton Barracks were of the usual pre-1939 German type but in much better condition than those at Fallingbostel. Most of the families could be accommodated in the town, although a few had to live in Brunswick which was reasonably close to Wolfenbüttel; as were the Harz mountains, with their ski slopes in winter and beauty spots for visiting in the summer.

Lieutenant-Colonel Morris has also described 1975 as 'Togetherness Year' after the scattering of the Regiment between Cyprus and Hong Kong that preceded it. There was now the opportunity to revive the Babington Shield for competition between the Squadrons and to bring the Canoeing and Fencing teams up to the mark. In 1975 the Regiment won the Inter-Regimental, BAOR and Army competitions for

canoeing, came third in the Army fencing finals, won the RAC
cup for skiing, and was only narrowly beaten in the Cavalry Cup
for football. It also had the distinction of being the only
regiment in the Royal Armoured Corps which was overstrength.
Much of the credit for this must go to the RSM, John Constable,
who, while the Regiment was abroad, devoted his time at
Tidworth to organizing the recruiting drives in Staffordshire.

1975 also saw the close of Colonel Anthony Bullivant's long
connection with the Regiment which had begun thirty-nine
years earlier when the 16th/5th Lancers were in India. A lot had
happened in between! He handed over as Colonel of the
Regiment to the author [Major-General James Lunt] who had
been the Commanding Officer from 1957-59 in Catterick.

The Regiment's role at Wolfenbüttel was armoured
reconnaissance. Armoured reconnaissance regiments were no
longer equipped with the wheeled armoured car. This had been
replaced by the Scorpion armoured and tracked reconnaissance
vehicle made by Alvis which had started to come off the
production line in 1973. It was the first vehicle of all aluminium
construction to be accepted into the British Army and was to
prove very popular. Armed with a 76 mm gun and a 7.62
machine gun mounted coaxially, it had a road speed of 50 mph
and was very versatile. One squadron was equipped with the
Scimitar which is similar in all but one respect to the Scorpion.
Its armament, however, is the 30 mm Rarden cannon with a
7.62 machine gun mounted coaxially. The Rarden cannon is
designed specifically for defeating the armour of light armoured
vehicles such as personnel carriers and is extremely accurate.
This range of vehicles was known as Combat Vehicles
Reconnaissance (Tracked) or CVR(T).

When a regiment is issued with new equipment it takes some
time, and involves intensive troop and squadron training, before
it can be mastered. Nevertheless the Regiment soon established
for itself an excellent reputation as an armoured reconnaissance
regiment in BAOR. But there were to be shocks in store. Early
in January, 1976, the Regiment was placed at 48 hours' notice
to move as the 'Spearhead' battalion for Northern Ireland. This
entailed infantry training for all. This was followed a few months

later by instructions that the 16th/5th Lancers would move in July on an 'Op Banner' (four months) tour in Northern Ireland where it would be operating in a purely infantry role. The families would remain in Wolfenbüttel in the care of a rear party.

It would appear to be the case that the worst part of the Regiment's tour in Northern Ireland was the rigorous infantry training they received beforehand at the hands of the Royal Green Jackets. 'They were pleasantly surprised at our standard of infantry skills and marksmanship,' we are told. 'They came with the idea that the cavalry had simply no idea and only appeared at the right place on the battlefield by pure chance or because they needed rations or more champagne. After their short time with us we introduced them to the complexities of Scorpion and they left fully converted ambassadors for the cavalry.' After a spell at Sennelager for a final polishing as infantrymen, the 16th/5th Lancers left for Northern Ireland at the end of July, 1976.

A Squadron was stationed at Portadown, a largely Protestant town where 'incidents' were few and far between; it had one Troop permanently on duty in the Maze prison. B Squadron manned the perimeter of the prison and had to be prepared to deal with disturbances within the prison if called upon by the prison staff. C Squadron was in Lurgan and was the only squadron to suffer casualties. Lieutenant Hepburn and Trooper Crutchley were wounded in the same incident when on patrol, Hepburn in the neck and Crutchley in the stomach and thigh. Fortunately both fully recovered after a spell in hospital. Despite this unfortunate event, the Regiment found the operational situation markedly quieter than it had been five years previously when stationed in Omagh.

Two other events in 1976 merit comment. In May the Regiment won the coveted Cavalry Cup at Football for the first time since 1932. They defeated the 15th/19th Hussars in the final. The other event was a change of Commanding Officers at the end of the year, Lieutenant-Colonel Vivian taking over from Lieutenant-Colonel Morris. This was noteworthy for being the first time an officer from outside the Regiment had been selected to command it since Lieutenant-Colonel Kaye in 1949. Vivian came from the 3rd Carabiniers which was in the process of being amalgamated to form the Royal Scots Dragoon Guards with the Royal Scots Greys.

1977 was the Silver Jubilee Year of Her Majesty The Queen and there was a spectacular parade by Rhine Army at Sennelager on 7 July to mark the occasion. The Queen reviewed the 4th Armoured Division, the Regiment being represented by 40 Scorpions and the Air Squadron's six helicopters. The organization of the parade which included infantry in their armoured personnel carriers, armoured regiments in Chieftain tanks, and the 16th/5th Lancers in their light tanks, not forgetting the Royal Artillery, the REME with their recovery vehicles etc., was extremely well done.

After reviewing the parade, the Queen mounted the reviewing stand where she was joined by Prince Philip and the President of the Federal Republic and Frau Scheel. There were 25,000 spectators, including the Colonels of all the Regiments and Corps represented on the parade. The drive past was led by E Battery, RHA, emphasizing the right of the Royal Horse Artillery when on parade with their guns to take the Right of the Line. They were followed by column after column of self-propelled guns, armoured personnel carriers, Chieftain tanks, recovery vehicles, ambulances and trucks. There were some 570 vehicles in all and since the tanks were moving at no more than 15 mph it was rather a lengthy business. The Regiment brought up the rear of the parade in their Scorpions, swinging past the reviewing stand at 40 mph, and dipping their guns in salute to The Queen, while overhead flew the Gazelle helicopters of the Air Squadron, their dressing as immaculate as the Scorpions. It certainly stole the show, the *Daily Telegraph* describing it as 'a modern-day charge', and Field-Marshal Sir Richard Hull as 'brilliant'. When a short time later the Colonel of the Regiment had an audience with the Queen, he was told how impressed she had been by the Regiment's performance.

As must have been evident to the reader so far, the British Army has to adapt itself to frequent reorganization. This may be due to changes in the strategic situation, or the introduction of new weapons, or to replacing conscription by voluntary recruiting. More often than not, however, the cause is economy, i.e. the Government's intention to cut back on the cost of Defence. This was the situation in 1976-77 when Britain was in the grip of an economic crisis. It led to what was called at the time the 'Restructuring' of the Army. Among other changes Brigades became known as Field Forces (it was not long before Field Forces became Brigades again) and infantry battalions lost their reconnaissance platoons for close reconnaissance; this now became the responsibility of armoured reconnaissance regiments, in addition of course to medium reconnaissance. The Regiment lost its Air Squadron, which became part of the Army Air Corps, its Assault (or infantry) troop in every squadron, and also its anti-tank guided weapon capability. B Squadron, equipped with Scimitars, provided close reconnaissance for several infantry battalion groups, each Troop accompanying its parent battalion wherever it happened to be serving, e.g. in Belize, Northern Ireland, or if training in Kenya or Canada. This provided Troop Leaders with a great deal of responsibility but it meant of course that neither the Commanding Officer nor, for that matter, the Squadron Leader saw much of the squadron as a sub-unit. A and C Squadrons, equipped with Scorpions, provided medium reconnaissance for the 4th Armoured Division.

It would be a bold, or perhaps even a foolish man who would claim that as a result of these changes the Regiment was as well organized as previously to carry out its role of reconnaissance. Undoubtedly it was not. However, the reductions in manpower spread across the board almost certainly avoided the disbandment or amalgamation of regiments, and to that extent were accepted as a necessary restructuring of the Army. It certainly had no effect on either the Regiment's morale, nor for that matter on its efficiency as a reconnaissance regiment, as is evident from the report made on the Regiment by the GOC 4th Armoured Division: 'The Regiment is fortunate in having its operational task on its doorstep and I am impressed by the very positive way in which they tackle it. This impression is reinforced by an excellent report on their Operational Test in which special praise is made of the Regiment's speed of assembly, speed of deployment, awareness and readiness of all ranks and vehicle availability. The 4th Division has gained an enthusiastic and professional reconnaissance force in the 16th/5th The Queen's Royal Lancers. They are fit, alert, well administered and clearly fit for role.'

That this was no flash in the pan was made clear at the end of 1 (British) Corps Exercise CRUSADER in October, 1980. This was the biggest exercise with troops carried out in BAOR for many years; it may well turn out to have been the last of its kind since there was a considerable amount of adverse comment in the West German press, due to the inconvenience caused to the civilian population. 4th Armoured Division acted as the 'enemy', and after the exercise was ended both the GOC and the Brigade Commander told the Colonel of the Regiment 'that the Regiment had been quite outstanding in Exercise CRUSADER and that there was no doubt in their minds that the 16th/5th Lancers were the best reconnaissance regiment in the Army'.

A solider of the 1980s

This was all the more to the credit of all ranks since the Regiment had spent from March to July in an infantry role in Northern Ireland. B Squadron was not eligible to accompany the Regiment because several of its troops had already carried out four-month tours in the Province with the infantry battalions to which they were attached for reconnaissance purposes. The Regiment was therefore brought up to strength by the attachment of 'M' Anti-Tank Battery RHA and 64 Amphibious Engineer Squadron, RE, a curious amalgam of all arms which worked extremely well. Conditions were quieter than they had been for some time, and although the Regiment was split up into independent troops all over the Province, there were very few incidents. Nevertheless, it resulted in no less than three Mentions in Despatches and the same number of GOC's Commendations. Command of the Regiment had changed prior to the Regiment moving to Northern Ireland, Lieutenant-Colonel Charles Radford having taken over from Lieutenant-Colonel Vivian, and the Colonelcy of the Regiment also changed about the same time, Colonel Henry Brooke taking over from Major-General James Lunt.

As part of the reorganization within the Army which took place in 1976-77 it had been decided that each regiment in the Royal Armoured Corps would take it in turn to be 'housekeeper' at the RAC Centre at Bovington for a two-year tour. This would not only make a sizeable reduction in manpower but would probably keep a regiment, or regiments, in being which might otherwise have to be disbanded or amalgamated. The Regiment was informed that it would take over this role at the end of November, 1980, having handed over at Wolfenbüttel to the 2nd Royal Tanks.

The role was hardly an exciting one. It entailed looking after the tanks and vehicles of the D & M and Gunnery Wings of the RAC Centre, and in carrying out the administrative duties inseparable from one of the largest training establishments in the British Army. A Squadron looked after the Gunnery Wing at Lulworth; B Squadron the D & M Wing at Bovington, where RHQ and HQ Squadron were also located. C Squadron was luckier. It was based at Tidworth to provide armoured reconnaissance for AMF(L) – the Allied Command Europe Mobile Force; this took it all over Europe from Norway to Turkey and proved to be an invaluable experience working with troops of so many different nations. While the Regiment was at Bovington Lieutenant-Colonel Radford also took the opportunity to widen the experience of his junior officers, sending them off to regiments in Cyprus, Hong Kong, Belize and Northern Ireland. One of them spent six months in Zimbabwe, and another with the King's Troop, RHA.

Belize, formerly British Honduras but no longer a colony, still depends on Britain for its defence against Guatemala, which lays claim to Belize. A reconnaissance troop of the RAC forms part of the garrison of a battalion group, each regiment in turn providing the troop for a six-month unaccompanied tour. Although Belize does not provide many amenities, it does provide a welcome change from service at home or in BAOR, but the hot and humid climate is not to everyone's taste. The territory is about the size of Wales, mountainous and densely forested. 'For the man who enjoys sailing, wind surfing, canoeing or diving there is a lot of fun to be had,' we are told. 'The sea is warm and the sun is usually shining... if it is not shining it will be bucketing down with rain.' The 16th/5th Lancers on several occasions provided a troop for the garrison which most people seem to enjoy, although, as one soldier has commented, 'Six months is quite long enough'. One military historian writing in the late 1960s wondered how the British Army would manage now that there was no longer an empire to be defended. He thought service in West Germany and at home would not compare with India, Burma, Egypt, Malaya, etc. Of course they do not, but it is surprising how widely travelled the modern soldier is. Air travel has made it possible for him to train in Canada, Kenya and other countries overseas, and many former British dependencies rely on the British Army to provide them with Training Missions.

The two years at Bovington passed remarkably quickly and in 1983 the Regiment moved to Tidworth in its former role of armoured reconnaissance, this time for the United Kingdom Mobile Force. At the same time Lieutenant-Colonel Wright

succeeded Lieutenant-Colonel Charles Radford as Commanding Officer. It fell to him to plan and conduct the imaginative parade on 15 July, 1983, when the Queen presented the Regiment with its new Guidon, and when the entire Regiment with its families and guests sat down to lunch in a huge marquee in the company of the Queen, our Colonel-in-Chief. It was a memorable occasion. The Queen obviously enjoyed herself, sending the following message on her return to the Palace: 'For Lt-Col J. A. Wright, MBE. I should like to congratulate My Regiment on a splendid parade, each man showed a precision, turn out and drill of the highest order, in very demanding conditions. I much enjoyed my day with you all. Elizabeth R.'

Tidworth was a familiar garrison town for the Regiment. Situated on the edge of Salisbury Plain, it provided considerable scope for the training of an armoured reconnaissance regiment. The Regiment had served there three times since 1945, and once between the two world wars. It was there on 5 July, 1926, that the late King Alfonso of Spain visited the Regiment as Colonel-in-Chief, playing as Back in the regimental polo team that afternoon in a match against the 7th Hussars, which the Regiment won by 5 goals to 4. Sixty-one years later, on the occasion of a State Visit to Britain by King Juan Carlos, King Alfonso's grandson, and Queen Sophia, the Regiment provided a street-lining contingent at Windsor. The Colonel of the Regiment, Brigadier Pownall, and Mrs Pownall attended the State banquet given at Windsor Castle by the Queen in honour of the King and Queen of Spain. The Regiment's connection with the Spanish Royal Family is now nearly 100 years old.

When Lieutenant-Colonel Wright handed over command in 1985 to Lieutenant-Colonel Derek O'Callaghan in October, 1985, (at the same time as Colonel Brooke handed over to Brigadier Pownall as Colonel of the Regiment), Wright recorded that he was handing over 'a living and a thriving family – our regiment'. It is probably true that this is how most officers and soldiers in the British Army regard their regiments, and it is certainly one of the strongest arguments in favour of the regimental system which has now been in existence for three centuries. From time to time efforts have

Colonel-in-Chief inspecting the Regiment on the Guidon Parade 1983

been made to change it on the grounds of administrative efficiency, but it will be a sad day if this ever happens.

Lieutenant-Colonel O'Callaghan joined the Regiment from the Queen's Royal Irish Hussars and it was during his year in command that the old Guidon, presented in 1959, was laid up in the WOs' & Sergeants' Mess. This was a break with the tradition that old Colours, Guidons and Standards are laid up in cathedrals and churches as consecrated objects. But Colonel Brooke obtained the Queen's permission that in this case an exception could be made. The Guidon was therefore handed over to RSM Lucas for safe and honourable keeping in October, 1985, by the author, on behalf of the Queen.

Lieutenant-Colonel O'Callaghan resigned his appointment in July, 1986, when he was succeeded by Lieutenant-Colonel Mark Radford, whose brother Charles had commanded the Regiment from 1979-82. This has set up something of a family record where the 16th/5th Lancers are concerned, nothing similar having occurred in the 300 years of the Regiment's history. By the time he assumed command it was known that the Regiment would leave Tidworth in November, 1986, for Herford in West Germany where it would be the armoured reconnaissance regiment in 4th Armoured Division, BAOR. Before that, however, the Regiment won the Cavalry Cup at football which somehow set a seal on what had been an extremely successful three years in Tidworth.

The life of the ordinary soldier serving in British Army of the Rhine nowadays has changed a great deal since the Regiment occupied Athlone barracks in Sennelager nearly forty years ago. Then it was an army of occupation, whereas today it is serving alongside a German army in NATO. Many of the privileges the soldier once enjoyed have disappeared, and he now has to compete on almost equal terms with the civilian population of the most prosperous country in Europe. In 1984 the Local Overseas Allowance was drastically reduced, whereas there was a time when the soldier was much better off than he is today. Moreover, field training has been much reduced, both in scope and in duration. This is due partly to economy and partly to objections raised when large areas of countryside are closed for manoeuvres. Again, partly for financial reasons, and partly to reduce turbulence, units remain in the same garrisons for six years or more. Since the soldier's life has always been one of frequent changes of garrison, the advantages of such an arrangement are probably offset by the disadvantages. Too long a stay in one place can lead to boredom. Even worse, restrictions on training probably means that too much time has to be spent on ordinary administrative duties, maintenance of vehicles, etc, which again leads to boredom. Finally, the astonishing changes which have taken place in Eastern Europe, including of course in East Germany, are bound to have their repercussions in BAOR, as every soldier knows. After nearly 45 years of stability, however menacing the 'Cold War', no one can foresee accurately what lies ahead.

When the Regiment arrived in Herford in November, 1986, no one, either in the Regiment or anywhere else in Europe or the world, could possibly have anticipated the changes that were to occur three years later.

Where the Regiment was concerned, the first priority was to settle into the normal routine of BAOR. The next was the requirement to commemorate in suitable fashion the three-hundredth anniversary of the raising of the 5th Royal Irish Lancers in June, 1689. Three hundred years of existence is, after all, something that deserves celebration, the British Army being in a sense unique among the world's armies, since throughout its history there has been no cataclysmic event like the French or Russian revolutions, nor the break-up of the Austro-Hungarian, German and Ottoman empires, to draw a line under the existence of regiments which in more than one instance would have been as old, if not older, than their British counterparts. 'Wynne's Irish Dragoons' had had their ups and downs, but they were still here 300 years after their original formation in the towns and villages of Enniskillen. It certainly merited suitable recognition. It was Lieutenant-Colonel Mark Radford who set in motion preparations for the great event, but it was Lieutenant-Colonel Scott, who succeeded him as Commanding Officer in March, 1989, who had to bring them to fruition. The precedents of previous and similar events during the past 20 years were disturbing. When the bicentenary of the raising of the 16th Light Dragoons was celebrated in 1959 the Queen had presented the Regiment with its first Guidon, following that by dining and dancing with the officers that same night. The great success of the presentation in 1983 of the second Guidon has already been told in a previous chapter. It would be difficult to repeat two such triumphs for the third time running – and this time without the Queen! As it happened, however, it turned out to be equally successful.

The Royal Warrant issued by King William III that authorized the raising of a Regiment of Dragoons 'out of our Inniskilling forces,' was dated 1 January, 1689. It was promulgated on 20 June, 1689, when Captain James Wynne, 'a gentleman of Ireland', was promoted Colonel to command Wynne's (Enniskillen) Dragoons, which later underwent several changes of title, viz: The Royal Dragoons of Ireland, the 5th Royal Irish Dragoons, the 5th Royal Irish Lancers, 16th/5th Lancers, and finally, the 16th/5th The Queen's Royal Lancers. It was thought appropriate that the tercentenary should be celebrated as near the original promulgation date as possible. This turned out to be the weekend of 22-26 June at Herford in the Federal German Republic in 1989.

Some 250 Old Comrades were present, all of them later loud in their praises for the administrative arrangements made for their comfort. There was a pageant depicting the history of the 5th Royal Irish Lancers from the original raising of 'Wynne's Irish Dragoons' until the 1922 amalgamation with the 16th Lancers. Many of the veterans who were present were astonished by the displays of the many activities in which their Regiment was involved. Things had indeed changed since their days! There was, of course, a formal parade at which the salute was taken by the Colonel of the Regiment, Brigadier John Pownall, who also presented the Regiment with a Commemorative Plaque on behalf of the Old Comrades. Messages were exchanged with the Queen, our Colonel-in-Chief, and in her reply the Queen said:-

'Please convey my sincere thanks to the Old Comrades, their wives and serving members of the Regiment and their families for their kind message of loyal greetings, on the occasion to mark the raising of the 5th Royal Irish Lancers 300 years ago. As your Colonel-in-Chief, I much appreciated this message of loyalty and send you all my very good wishes for a very enjoyable Regimental weekend.'

This will be an appropriate moment to end this chapter's account of the Regiment's activities between 1968 and 1990, and to look in more detail at the Regiment's experiences in Aden, Cyprus, Beirut, and most recently in the Gulf War in January-February, 1991.

CHAPTER ELEVEN

Hong Kong & Aden
1964

1964 was the year when the Regiment was probably more widely dispersed than at any previous time in its history. Regimental Headquarters, Headquarters Squadron, and one Sabre Squadron were in Aden; one Sabre Squadron was in the Gulf, half on shore in Bahrein, and half embarked in a Tank Landing Ship (LST) which was crewed by the Royal Navy. The third Sabre Squadron was in Hong Kong. All three Sabre Squadrons were equipped with the Centurion tank, upgunned with the 105 mm. The Reconnaissance Troop (in Aden) had Ferret scout cars. The Commanding Officer, Lieutenant Colonel Holland, when not airborne visiting his far-flung squadrons, was based in Aden. He answered to several masters, among them the GOC Middle East Land Forces (for Aden and the Gulf), the Commander British Forces (for Hong Kong), and the Commander 48th Gurkha Infantry Brigade (also for Hong Kong). Several other senior officers had some claim on his attention, such as the Commander British Troops (Bahrein) and the Commander Federal Regular Army (FRA) in Aden. In this particular instance it is probably true to say that safety lay in numbers.

Hong Kong was then still an old-style British colony governed with due attention to the protocol that had distinguished British rule in India. The Commander British Forces was a Lieutenant-General, unique among his colleagues in being entitled to be addressed as 'His Excellency'. The Major-General

Captain Julian Johnston's tank driving past His Excellency Sir David Trench. Queen's Birthday Parade, Hong Kong 1964

Brigade of Gurkhas also had his headquarters in Hong Kong, the garrison being provided predominantly by the Gurkha Rifles. It was one of the last of the imperial garrisons and therefore top of the list for those senior officers in the Ministry of Defence who found it a compelling necessity to get away from Whitehall in order to visit the troops. Each one of them had to be welcomed, briefed and entertained. Ceremonial played an important part in the life of the British Army in Hong Kong, as A Squadron (Major Pownall) was soon to discover.

A Squadron formed part of 48th Gurkha Infantry Brigade, and no one who has served with Gurkhas can have failed to be impressed by them as soldiers. This went a long way towards compensating for the sorry state of Sek Kong camp in the New Territories, often short of water and besieged by swarms of mosquitoes from the surrounding rice fields. The tanks were not in their first youth but their mere presence impressed the

Guard of Honour for Prince Sihanouk of Cambodia. One of many provided by A Squadron in Hong Kong

locals. It was touch and go, however, on the annual Queen's Birthday parade if they would get through the parade without breakdowns.

The training areas were small and inadequate for armour, but there were observation posts along the border with China which the Squadron manned, as well as carrying out border patrols on their feet. There was plenty of sport, sailing being particularly popular, and Kowloon did not lack for attractions of another kind provided one could afford them. The Hong Kong climate was bearable for much of the year, but the typhoons added to the hazards; there were no less than five during A Squadron's tour, the winds sometimes gusting to 120 mph. The fact that the Squadron could have their families in Hong Kong gave it an advantage over the other squadrons and provided a welcome boost to morale.

Aden was a different kettle of fish altogether. It, too, was a colony (or rather had been until 1962), but it bore no comparison with Hong Kong, for climate, amenities or prosperity. 'At first sight,' wrote one visitor, 'Aden strikes most newcomers as unmistakably the most repellent city they have ever set eyes on.' Nor did it improve on closer acquaintance. The climate is hot and humid, dust storms occur regularly during the summer, and the local population was not particularly friendly. At one stage in British imperial history, when Aden was controlled from India, it was used as a punishment station for British units that had misbehaved elsewhere. The Colony itself covered little more than 75 square miles, beyond which there was tribal territory where visitors were unwelcome, every man carried a rifle, and the blood feud flourished. Between the borders of the colony and the high mountains of the Yemen there were still areas into which Europeans had seldom penetrated. One of them, the Radfan, was only 70 miles from Aden as the crow flies, but little attempt had been made to administer it. The tribes who lived there, among whom the Quteibi were the most numerous, were wild, fanatical, treacherous, and miserably poor and under-nourished. Outside the colony, 'Up Country' as it was known in Aden, there were hardly any metalled roads, only dirt tracks linking the scattered villages. It was a land where the Arabic saying 'The enemy of my enemy is my friend' had a real meaning. This wild, mountainous, craggy and desolate region had a certain charm of its own, particularly in winter when the heat and thirst of summer was only a memory, but it was safer to fly over it in a plane than to traverse it on foot, alone and unarmed. Since 1958 it had been known as the Federation of South Arabia.

Britain annexed Aden in 1839 because of its strategic situation on the long line of communications with India. It was governed as a fortress and after the Suez Canal was opened Aden became one of the most important coal-bunkering ports in the world. In 1954, after British Petroleum had built an oil refinery there to replace Abadan in the Gulf, Aden became the second biggest oil-bunkering port in the world — after New York. Despite all this, the British had not sought from the outset of their occupation to extend their rule much beyond the fortifications of Aden itself. They left the numerous tribes, with their feuding sultans, sheikhs and amirs, to fight among themselves, bribing them to keep the peace by gifts of rifles and ammunition. It was only in 1958 that they cajoled them into a loose kind of Federation, which they persuaded Aden Colony to join in 1962. It came to a sorry end in 1967.

Aden's garrison had been mostly supplied by the Indian Army until 1937 when the Colonial Office assumed responsibility for it. From 1928 the Royal Air Force was responsible for the Colony's defence, including keeping the tribes quiet 'Up-Country'. A locally enlisted force, the Aden Protectorate Levies, was responsible for protecting the numerous landing fields. The policy adopted by the RAF was called 'Air Control'. An offending tribe would be told that it must cease its lawless activities and provide hostages for its continued good behaviour. If it failed to do this its tribal territory would be proscribed and would be liable to attack from the air, which included bombing. The bombs were small, the mountains were honeycombed with caves, and the majority of the casualties were camels, goats and donkeys. The mud-brick houses might be knocked down but could easily be rebuilt. However, the tribes found it very inconvenient to leave their homes and fields and live in caves. They found it preferable to submit to the terms of government. For much of the time between the two world wars the RAF maintained the peace by this method over wide areas of the Middle East. The main drawback was that it absolved the government from having to open up the country by building roads and keeping them free for peaceful traffic. In more than a century's rule in South Arabia, Britain constructed hardly a yard of tarmac outside Aden itself. But apart from this, Air Control had shown itself to be an economical and effective way of maintaining order among the wild and lawless tribes of Aden's hinterland.

No one was much concerned by the ethics of bombing tribesmen during the 1920s and 1930s, but it was a different story after the Second World War when entire cities had been levelled to the ground by bombing. HMG came under criticism for employing Air Control and the last time it was used in Aden was in 1961, when it was very effective. However, when the High Commissioner, Sir Kennedy Trevaskis, at the end of 1963 asked for RAF support to evict some dissident tribesmen who had taken refuge in the Radfan, it was refused. Instead, it was suggested that some kind of 'minor' land operation should be mounted instead. This could be carried out by the local Arab Force, the Federal Regular Army (FRA), which the author was commanding at the time.* The aim of the operation, if it could be said to have had an aim, was to drive a handful of dissident tribesmen out of the Radfan and to demonstrate the government's ability to enter and leave the Radfan at any time it chose to do so. For this purpose a motorable track was to be built up the Wadi Rabwah into the Radfan proper. A sizeable force was assembled, mostly Arab in composition, and Operation 'Nutcracker', as it was codenamed, began on 4 January, 1964. The codename was chosen by the author because we were using a sledgehammer to crack a nut. It would have been cheaper, and far more effective, to have bribed the local tribes to drive out the dissidents themselves.

The Radfan was a massif of considerable extent, and was dominated by the 5,500 ft peak, Jebel Huriyah, from where the lights of Aden, 70 miles away, were clearly visible at night. Despite this proximity to Aden, little or no attempt had been made to penetrate the Radfan, apart from the Wadi Misrah which was the territory of the Quteibi tribe, as untrustworthy and as fanatical a tribe as any to be found in Aden's hinterland. One example of this isolation was the author's experience on the second day of the operation when he accompanied a platoon that was landed on the Bakri ridge, which was to acquire some notoriety during the second phase of the campaign. As the Arab soldiers clambered out of the helicopter a few of the tribesmen, both young and old, gathered to watch the proceedings. Squatting on their hunkers, their cheeks bulging with *Qat* (a local plant that has narcotic effects, induces a violent thirst and tastes of privet), they looked on peacefully although several were armed. One of them, much older than the rest, fixed the author with the stare of the Ancient Mariner

---

* I was strongly opposed to the Radfan operation, as were the Federal ministers of Defence and Internal Security. I believed it would escalate beyond the ability of the FRA to control, as eventually proved to be the case.

until asked why he was doing so. 'I've seen him before,' the old man replied. 'His name is Yacov.' 'He's the only *Ingleezi* I've ever seen,' he added. Lieutenant-Colonel H. F. Jacob (*Yacov* in Arabic) was a political officer in Aden from 1904-07 and during that time he did enter the Radfan. The old man who confused him with the author must have been a small boy at the time. He had apparently never left the Radfan although Aden was so close. It serves to show the extraordinary isolation of some of the tribes in their mountain fastnesses.*

The assembled force for Operation Nutcracker did experience some opposition, but it did not amount to much. A great deal of effort was devoted to constructing the track up the Wadi Rabwah (by British Sappers) but it was quickly destroyed by the tribesmen after the force had withdrawn. A troop of Centurion tanks was sent up to the Radfan from B Squadron shortly after it had arrived in Aden. It was hardly tank country but the aim was more psychological than military; it was hoped to demonstrate to the tribes that the government had a great deal of military force at its disposal. The fact that the troop happened to be commanded by the author's son aroused a certain amount of interest, notably in the *News of the World*, whose correspondent happened to be visiting the Radfan. This led some wags in the Regiment to describe the Radfan as the 'Lunts' Front'. 1st Troop did not have the opportunity to fire its guns in anger, although it did take part in an impressive fire-power demonstration, leading one of the watching tribesmen to tell the author, 'You wait until we get tanks!'

As had been forecast, the scope of the operations rapidly escalated. There were large numbers of dissidents taking refuge in the Yemen, then in the throes of revolution, and they seized the opportunity to come down to the Radfan, well armed and equipped by the Egyptian army then operating in the Yemen. These dissidents were provided with an excellent jumping-off spot for operations in Aden itself, and there can be no doubt that the terrorist activities already taking place in a small way in Aden itself received a considerable boost from the decision to open up a campaign in the Radfan on Aden's doorstep. By the end of January, however, Operation Nutcracker's aim had

* Many tribesmen, however, did travel far afield before returning to their remote valleys. Many worked on building sites in Birmingham and in at least two instances returned with English wives to their mountain villages.

"Up Country" – Aden 1964

been achieved and the majority of the force was dispersed.

However, it proved to be only a lull before the storm. Large numbers of dissident tribesmen infiltrated the Radfan from the Yemen, bringing with them modern weapons, ammunition and mines. Incidents increased until, by April, it became necessary to reopen the campaign, which had escalated beyond the capacity of the FRA to handle it. British troops, under a British headquarters, took over from the FRA, although at least one FRA battalion was placed under British command. The second phase of the Radfan operation proved to be a great deal tougher than Operation Nutcracker. It cost a lot more money and resulted in many more casualties (on both sides) than the earlier phase. By the time it was finished at the end of August there was nothing to show for becoming involved in the first place.

Only the Reconnaissance Troop and one troop from C Squadron were involved in the second phase. The curtain-raiser for the Recce Troop was the mutiny of the King's African Rifles in Tanganyika and Zanzibar in January, 1964. Part of the Troop was embarked on *HMS Centaur*, landing at Dar-es-Salaam on 24

January. Captain Michael Brooke, who was commanding the Troop, claimed that this was the first time the Regiment had participated in an amphibious landing, but he was wrong. Two troops of the 16th Light Dragoons took part in the capture of Belle Ile on 25 April, 1761, losing one of the troop leaders in the process. Brooke was luckier. There was, in fact, no fighting. After changing ships twice more, the Recce Troop returned to Aden at the end of February.

It had hardly settled down before it was sent up to the Radfan where it remained for nearly six months, for most of the time under Captain Gerard Henry. It operated under the command of 4th Royal Tanks, who were equipped as an armoured car regiment. The troop was kept busy escorting convoys, keeping open the dirt road from Aden to Dhala on the Yemen frontier, and similar activities. They were not without danger since the enemy were well supplied with anti-tank mines and knew how to use them; both the Brigade Major and DAA & QMG of the FRA lost their lives in a mining incident early in April, and there were other casualties. There was always the risk of ambush when venturing up the narrow wadis, with the tribesmen as sure-footed as goats amidst the crags and boulders above. It was stiflingly hot at night as well as day, and sand got into everything, including a mug just filled with tea. It was really hard soldiering but in the opinion of every man in the Troop a thousand times better than Aden.

Major (then Lieutenant) Edward Curtis was sent from A Squadron in Hong Kong to relieve Captain Henry in temporary command of the Recce Troop and has provided an interesting account of its activities:

'I was sent to Aden from July to September, 1964, to command Recce Troop while Gerard Henry was away. Having spent a week at Falaise Camp learning about scout cars, tactics and the Browning machine gun, the Troop, consisting of six scout cars, was sent up-country to Thumeir, about 70 miles from Aden. Thumeir was the main base for operations in the Radfan. There we were placed under command of a 4 RTR Squadron Leader and told that the Troop would spend about a

Landing from Sir Lancelot

Troop Leader, Lieutenant Edward Curtis, in Ferret Scout Car as his troop returns from ten days in Wadi Taym

week at a time in each of three areas — Thumeir, Wadi Taym and Dhala.

'We drove out early each morning from Thumeir to recce the various dirt tracks in case mines had been planted overnight. The chances of actually spotting a mine were slight and the only way to locate one was to run over it, losing a scout car wheel in the process! We also escorted convoys of trucks and other vehicles carrying supplies or personnel to the Wadi Taym or to various other locations manned by the Marines, 1 Royal Anglians and the Royal Artillery.

'Wadi Taym was very much the sharp end of the action at this time and was controlled by the Royal Anglians. There were three Company posts, Bn HQ, and an Artillery battery in the Wadi. We were required to have two scout cars with each of the Companies for mine patrolling, escort duties, occasional daylight sorties, and to stiffen up the company position when fired on. I took my patrol to the Company farthest up the Wadi and stayed with them. Daily at about 4 pm the dissidents machine-gunned the camp somewhat inaccurately. It was very difficult to locate them.

'One day, however, I went out with another scout car shortly before 4 pm to see if we could surprise the Arabs while they were taking up their position. We were about half a mile from the Company position and looking at a cave across a small wadi when there was a loud bang! I peeped out of the top of the scout car and looked to either side when there was yet another bang. Something whistled overhead and I saw it land and explode near some disused buildings to the left of my vehicles. I assumed that someone was trying to bazooka us and must be located to our right. Then I was immediately distracted by machine-gun bullets zinging off the scout car and by a blinding flash in the small periscope in the front of my vehicle. Besides being bazooka'd from the right, I was also being machine-gunned from the front! I sent a message to the Company and opened up with the Browning at all the likely spots.

'The action lasted about ten minutes, we remaining where we were and hoping our armour would protect us, which of

course it did. It was impossible to locate the enemy. The
Anglians quickly sent out a Platoon to assist us and search the
area but they found nothing. The enemy had covered their
tracks most efficiently. After that incident we continued to be
sniped at long range on a daily basis but imagine we succeeded
in the aim of preventing ammunition in quantity being
transported down the Wadi Taym.

'From Wadi Taym the whole Troop was sent to Dhala to
escort part of an FRA battalion, and also to strengthen the Dhala
defences. We were sent out on daily mine patrols and also took
part in two cordon and search operations. The main excitement
came one night when there was another loud explosion which
we discovered subsequently was made by a bazooka being fired
at the Political Officer's house. This was followed by a stream
of machine-gun bullets whizzing over our camp. We raced from
our beds to our scout cars with bullets flying everywhere.
However, it was difficult to reply to the fire without plastering
the Amir's palace. Then the Arab CO of the FRA battalion
jumped on the back of my scout car and directed my fire on
to some likely targets. The action lasted about 15 minutes. I
remember being amazed by the response of the Royal Marine
Commando in their adjacent camp. They opened up with
everything, mortars, machine guns, rifles, but luckily their
direction of fire was safer than ours. The firework display was
most impressive. A few days later I returned with the Troop to
Thumeir where Gerard Henry was waiting to reclaim his Troop,
worse luck!

'However, I cannot end this short but interesting experience
in the Radfan without mention of Cpl (now SSM) Wells. He was
a bright young soldier who displayed great skill, sound sense
and courage. He was awarded a GOC's Commendation. I
believe he was involved in an attack on a village by 3 Para. One
of the Paras was badly wounded and Cpl Wells shielded him
with his scout car, and indeed with his body for a short time,
while the bullets rained down from the village. The wounded
Para was saved, the scout car was riddled with holes, and yet
Cpl Wells was able to give the Paras accurate covering fire for
their assault, which proved to be successful.'

B Squadron, having had the vicarious satisfaction of sending one of its Troops up to the Radfan, changed over with C Squadron in Bahrein during February, 1964. It was a curious role for a cavalry regiment, drifting round the Gulf in a LST, and enduring climatic conditions that resembled a sauna. Those not at sea did not find much to amuse them in Bahrein, apart from making polite calls on the Ruler, who was lavish in his gifts of gold wrist watches. On C Squadron's return to Aden, it too was required to send a Troop of tanks up to the Radfan. It was 4th Troop and it had a rather more exciting time than 1st Troop of B Squadron had had in January. It took part in an advance down Wadi Misrah, stronghold of the Quteibi tribe, firing its guns in anger on several occasions, and being fired upon in return. There were no casualties in the Troop but it was the first time tanks in the British Army had been in action since Korea. A solo contribution to the Radfan campaign was made by Captain Flynn who was Forward Air Controller with 45 Royal Marine Commando. This enabled him to grow a beard and wear out his boots on the rocky hillsides; fortunately he was well used to walking, having served in the Royal Ulster Rifles before joining the Regiment.

Many in the Regiment serving in Aden, whether or not they had been anywhere near the Radfan, received the General Service Medal with the 'Radfan' Clasp. This was denied to all those who had begun the campaign in January under FRA command. No one has been able to explain why this differentiation was made. Although the Regiment's part in the campaign had been a modest one, it does, however, allow it to claim that it was involved in the last old-style Colonial campaign of the British Empire, the kind of campaign to which the British Army had so long been accustomed, against the Afghans, the Pathans, the Zulus and so many others. It was indeed the end of an era.

CHAPTER TWELVE

Northern Ireland 1971-73

DESPITE the Regiment's partly Irish ancestry, it had not crossed the Irish Sea for nearly sixty years until it arrived in Omagh in County Tyrone on 5 November, 1971. For more than a decade prior to 1969 Northern Ireland had been a military backwater, but in that year the long-smouldering quarrel between the Northern Irish Catholics and Protestants burst into flames, and the fire has yet to be quenched more than twenty years on. The internal security situation in the Province is one of the British Army's principal commitments, and when the Regiment arrived in Omagh the security situation was going from bad to worse. It had not noticeably improved by 1973 when the Regiment's tour in Northern Ireland came to an end.

There was an unreality about the situation in Northern Ireland which is difficult to describe but which was part and parcel of every soldier's daily life when serving there. Unlike similar internal security situations elsewhere in the days of Empire, there was nothing to differentiate friend from foe — neither language, way of life, dress, colour, nor appearance. IRA gangs were indistinguishable from the rest of the population. Life in Northern Ireland, even in republican strongholds like the Falls Road in Belfast or in the Bogside in Londonderry, was very much the same as life in Walsall or Stoke. The people looked the same. There was nothing in their outward appearance to distinguish a Catholic from a Protestant. Until bitter experience had taught him otherwise, the ordinary friendly British soldier found it hard to comprehend that the ordinary friendly Irishman who stood him a drink at one moment in the local pub might shoot him down like a dog in an ambush on his way back to barracks.

Some of this puzzlement is well expressed in an extract from the 1972/73 edition of *Scarlet & Green*: 'It has often been said that the English have never understood the Irish and you certainly see some very puzzled soldiers in this country. How do you understand the genuinely charming people who give you cups of tea one day and throw bricks at you the next? What kind of double talk calls on "all people loyal to the Queen to consider the British Government and the Army as their enemies"? Gunman and bomber, both Protestant and Catholic, could not survive without the connivance of at least some part of the general public who profess only to want peace. Innocent men, women and children die daily in indiscriminate attacks whose justification it is impossible to define.'

One major issue which had to be faced was the presence of the families. Omagh was an 'accompanied' posting and, operational commitments notwithstanding, and naturally they had priority, some arrangement had to be devised whereby husbands saw something of their wives and children. The solution was a compromise involving rotation of squadrons through two fully operational weeks, followed by one week 'in reserve' in Omagh. The reserve squadron provided a stand-by troop, suitably reinforced and ready to turn out at 10 minutes' notice. The reserve squadron also met any unforeseen demands for extra help (there was never a week when this did not arise), devoted some time to training and maintenance, and took what rest and recreation was possible. At that early stage in what might be termed the Emergency in Northern Ireland, every effort was made to maintain an impression of normalcy, insofar as this was compatible with military efficiency. As one example of this the Regiment's Married Quarters were not enclosed by barbed wire, although the necessary material was held available to do so if required. This had to be done within two weeks of the Regiment's departure when the Married Quarters were mortared.

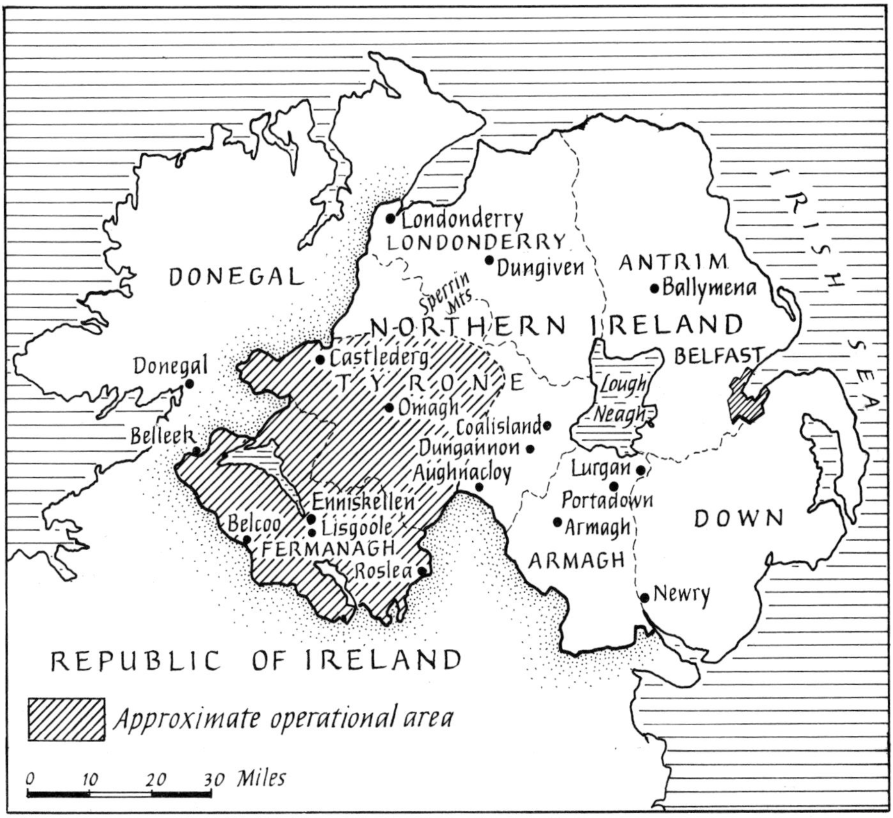
Londonderry
LONDONDERRY
Dungiven
ANTRIM
Ballymena
DONEGAL
Sperrin Mts
NORTHERN IRELAND
Castlederg
TYRONE
Omagh
Coalisland
Dungannon
Aughnacloy
Lough Neagh
BELFAST
Donegal
Belleek
Enniskillen
Lisgoole
Belcoo
FERMANAGH
Roslea
Lurgan
Portadown
Armagh
ARMAGH
DOWN
Newry
IRISH SEA
REPUBLIC OF IRELAND
Approximate operational area
0 10 20 30 Miles

The problem throughout lay in the balancing of risks against the need to maintain morale, not only within the Regiment, but also amongst our supporters in the civil population, who were numerous. It placed a heavy burden on the Commanding Officer, Lieutenant-Colonel Dennis, who could not be expected always to get it right. As but one example, off-duty soldiers were permitted to make use of local facilities for sport and other forms of recreation, but they obviously ran a risk while doing so. This was the case with Lance-Corporals Alder and Revitt who were wounded on 7 February, 1972, when returning to camp after visiting a pub. It was typical of the topsy-turviness that so many soldiers came to associate with Northern Ireland that after the wounding of Alder and Revitt the local newspaper, *The Tyrone Constitution*, launched a very successful appeal for them. Another example of the generosity

Rossville Street, Londonderry, late 1971

of ordinary folk was the cheque for more than £2,000 presented to the Commanding Officer by the people of Fermanagh for Christmas comforts. When set against the killings, wounding and maimings that occurred almost every day, it truly was puzzling.

In August, 1971, prior to the Regiment's arrival in Northern Ireland, a great many IRA supporters had been arrested and interned; about 326 were picked up. This seemed to open the floodgates to ever-increasing acts of terrorism both in Northern Ireland and on the mainland. At the time of internment there were 14 infantry battalions in the Province, but 'incidents' continued. There were 124 in November, and 98 in December, 1971. Two more infantry battalions and the equivalent of another armoured reconnaissance regiment had to be sent to Northern Ireland in November. Londonderry had become virtually a 'no-go' area to the security forces. There were no less than 723 explosions in the Province between August and December, 1971. In Omagh the Regiment had the 4th and 6th Battalions of the Ulster Defence Regiment (UDR) under operational control, as well as an infantry company on a four months' tour in Northern Ireland. *Scarlet & Green* was not far wrong in its comment: 'Some of us can still just remember a way of life called the BAOR training year, the main feature of which was comparatively short periods of intense activity, interspersed with long periods of military thumb-twiddling. Life in Northern Ireland is certainly different from that!'

Even Headquarters Squadron, considered normally by the Sabre squadrons as leading a sedentary life, had to take part in foot and mobile patrolling, arms searches, looking for land-mines, crowd control, and other measures intended to defeat the gunmen. Its first real action was in March, 1972, at Belleek where the RUC police station is only 100 yards from the border with the Republic. An intense exchange of fire developed as the Squadron's advance party arrived on the scene, effectively pinning down part of the echelon. There were no casualties but shortly afterwards Bandsman Stock was shot in the foot when on patrol. As Bandmaster Tomlinson pointed out, this was hardly the reason why men joined the Army as musicians.

Gradually and almost imperceptibly the pattern of operations changed. Though vehicles continued to be used, particularly for administrative purposes, they declined in importance as helicopters and flat feet came to dominate activities. In addition to the two main bases (Lisanelly Camp and St Angelo airfield), detachments were maintained at many RUC police stations, particularly on the border, where many attacks were made against the RUC; the first use in Ireland of the Russian-made RPG 7 anti-tank rocket launcher was on 28 November, 1972, against Belleek RUC station, resulting in the death of Constable Keyes.

Controlling the border with the Republic was a continuous and almost insuperable problem. Unlike the Berlin Wall and the Iron Curtain that so effectively corralled the people of East Germany until recently, the border with the Republic of Ireland was an open one. This gave the gunmen an excellent refuge whenever it became too hot for them in the Province. Many and various attempts were made to control, or close, crossing points, but without much effect, as the following account by

Major (now Colonel) Charles Radford, then commanding A Squadron, makes clear:

'One of the least successful tactics carried out in Northern Ireland in the early 1970s was that designed to close all unauthorized border crossing points. Any study of a map reveals a spider's web of crossings – roads, tracks and bridges weaving a random pattern across an ill-defined demarcation line. A protracted operation was launched with the aim of sealing all but approved crossing points, and so making the job of the security forces that much easier in controlling the border and the movements of the IRA terrorists to and from the Republic. It was all very well in theory, but in practice the engineering skills and determination of the local republican population were greatly underestimated, as were the legitimate needs of local farmers with land on both sides of the border.

'The story usually began at Lisgoole school, just south of Enniskillen, where the Fermanagh squadron was billeted in the early days on a rotating basis. The evening before the operation the Sappers would arrive with their equipment – light and medium wheeled tractors, a tracked digger on a low-loader, truck loads of concrete bollards and explosive, and the usual command and support vehicles. All squadrons in the Regiment will remember having laid on numerous border closure operations, at which they became extremely adept. Their drawback was that after a time they became fairly predictable once the operation got under way and offered an open invitation for some kind of confrontation with the gunmen.

'Before first light the convoy set off for the border. Painstaking attention to the order of march, and a painfully slow progress to cater for the Sappers, saw us on our way with excitement mounting in anticipation of a possible engagement with the IRA at some stage during the day. Shortly after it was first light the first task was to clear the immediate approach to, and exact location of, the border crossing. Assault troopers on foot cleared every yard, backed up by Saracens and Ferrets. One of our helicopters observed overhead.

'Once the area was sealed, the Sappers moved in to carry out their highly exposed task. We soon got to know the drill. There were firstly the beehive charges to blow chambers in the road, or track, for the main explosive charges. Hundreds of pounds of plastic explosives were slid down metal tubes into these chambers; no manuals were consulted here, resulting in over- rather than under-kill. It avoided the administrative problem of having to take explosives back with us. The ringmains were laid and a healthy safety distance was allowed. Competition for "pressing the button" was intense, and when "she went", it was an impressive sight. A rumbling, blasting boom lazily threw tons of earth and rock hundreds of feet into the air, to rain down around the 15 feet deep crater, not a few projectiles hitting closer-sited armoured vehicles with a resounding clang.

'It was usual at this stage that the opposition across the border moved up to have a go at us. One squadron leader was caught while inspecting the crater with the Sapper sergeant. They were deciding how best to deal with the Republican tricolour hoisted in a nearby tree. The rifle fire resulted in a hasty dive to the bottom of the crater, to the amusement of those looking on from the safety of their closed-down armoured and scout cars. The mere demolition seldom satisfied the Sappers and hours were spent with picks, shovels and dozing on "crater improvement". This was to make the obstacle as difficult to repair as possible. Sometimes streams were diverted into the crater to make a sizeable salmon pool.

'This stage of the operation — "Crater Improvement" — provided the gunmen with a wonderful opportunity. As unarmed Sappers widened the crater with their diggers and tractors protected only by a very synthetic armour, they were vulnerable to marksmen across the border. The Sapper drivers had often to dive from their cabs to safety below ground level. Despite such interruptions, the job would usually be completed by noon or the early afternoon. The Sappers would then load up prior to departure, their last act being the positioning of enormous concrete bollards on the road as an added deterrent.

'However, the local Republican population were ingenious in overcoming the most fearsome obstacles, and there was of course no possibility of the security forces remaining to guard the sites. Old and damaged cars were the primary repair material, together with ample supplies of turf and hard core, all in ready supply across the border. Local manpower was also available, keen and determined, and within a surprisingly short time − 24 to 48 hours − cars had been piled into the crater, crushed down and padded with hardcore and turf to make a wobbly but perfectly usable crossing.'

Inevitably this was claimed as a victory by the IRA. Occasionally troops were sent to prevent attempts to reopen the crossing but this only resulted in confrontation with the locals, leading to the firing of plastic bullets and the use of CS gas. It gave rise to local resentment and sometimes caused unnecessary minor casualties on both sides. It is therefore not surprising that such ineffective attempts to seal the border were eventually abandoned.

'The Regiment looked after a considerable stretch of the border,' writes Colonel Charles Radford. 'Much time and effort was expended in somewhat vain attempts to curb terrorist cross-border activities. The area is on the whole wild and remote country dominated by mountainous and boggy terrain that could quickly turn from the black and menacing to surprisingly beautiful when the sun lit up the purples, browns and greens of the upland moors. Road communications were for us slow and easily predictable while offering the terrorists countless crossing places across the border. There were loughs, rivers and bogs in abundance, hindering and channelling road movement for the security forces. And of course it regularly rained.

'Where small communities had sprung up at the main natural crossing points, there would usually be situated a small, standard-designed and fortified police station.... These outposts of empire were manned by a station sergeant and a handful of RUC constables. By 1971 normal policing was becoming increasingly difficult due to the steady increase in IRA activity, as still remains the case. It was standard practice to reinforce these outstations with a Troop, rotating on a weekly basis. Castlederg, Belleek, Belcoo, Roslea and Aughnacloy all had their fair share of terrorist activity, always when least expected. The normal daily and nightly routine consisted of foot and vehicle patrols — at that stage in the campaign usually by armoured cars — acting on information gleaned from the resident RUC, and increasingly from the Ulster Defence Regiment. Tangible success was hard to come by but from time to time the vehicle check-point, set up at random, would be rewarded by the detention of a suspected terrorist, or by the finding of arms, ammunition or explosives. It was time-consuming and tedious work carried out under the constant threat of attack from remotely detonated mines or small arms fire. Under current conditions most movement is now carried out by helicopter or on foot since vehicle movement in border areas is an open invitation to ambush. Many an armoured car returned from patrol with bullets in the tyres and in one case with armour-piercing rifle bullets three-quarters of the way through the armour of a Ferret scout car.... Lt Clive was amazingly lucky to survive a mine attack in his Ferret, the turret of which was found in an adjoining field some 25 yards away. Clive was also found in the field, having somehow cleared the hedge with nothing worse than a broken jaw and injured pride!

'Successful countermeasures against the remotely detonated mine were developed and huge quantities of home-made explosives (normally fertilizer mixed with diesel oil) were retrieved from culverts and cuttings by assault troopers laboriously checking on either side of roads and tracks.... All this involved the sealing off of areas while those most courageous of men, the ATOs (Ammunition Technical Officers), made the device safe and in most cases blew it up in an open field nearby.... In those early days cross-border police liaison was patchy, unpredictable and on the whole unproductive, there being no question of our being allowed a yard into the Republic to investigate the firing point of a particular explosive device.

'Despite our efforts, and those of the RUC and UDR, the IRA were able to carry out their murderous work in their own time and with little likelihood of immediate detection. It was the Protestant farmers in the remote border areas who became the targets of the sickeningly callous campaign of IRA murder. After

Joint operation with the Ulster Defence Regiment

becoming members of the UDR in order to defend both their livelihoods and their families, these gallant men were placed at even greater risk. With remarkable courage they would farm all day and go out on patrol most nights of the week. Their energy certainly matched their bravery. It was one of the more depressing tasks for the Regiment to send representatives, usually the Commanding Officer and Squadron Leader, to the funerals of so many brave and worthy comrades in the UDR.'

It was indeed a 'Twilight War'. 'The trouble about Northern Ireland,' one soldier has said, 'is that you need eyes in the back of your head, because you don't know who to trust.' It was not as if the British Army lacked experience of internal security operations; there had been plenty of opportunities to learn since 1945. But in some curious fashion Northern Ireland was different. 'Unfortunately we play the game under very strict rules,' reported *Scarlet & Green*. 'These give the opposition a very good chance of survival. It is like playing a football match where your players aren't allowed into the opposing half and cannot go for the ball until the other side has had a shot at goal. Even when you get the ball you can only take a long-range shot and then wait for the next attack to develop. The team supporters can change sides at the drop of a hat... and spend a lot of time fighting together on the field of play.'

The Regiment was not involved in the events in Londonderry on 31 January, 1972, which has gone down in history as 'Bloody Sunday'. It did, however, man a line of road checks along the Sperrin Mountains south-east of Londonderry. The Regiment was not on the 8th Infantry Brigade radio net and therefore had no idea of what was taking place in Londonderry. 'Memories are of a magical Irish day of hard frost, a light sprinkling of snow on the ground, bright sun and clear blue skies,' writes Lieutenant-Colonel Dennis. 'How different experiences can be when separated by such a short distance.' On the previous day A Squadron had had to prevent an illegal Civil Rights march from Dungannon to Coalisland. It was led by the formidable Miss Devlin. She had to be forcibly restrained from going any farther by Lance-Corporal Stone to whom she claimed that she was only taking a Saturday walk with friends!

A Squadron was also involved in Operation Motorman which took place on 31 July, 1972. It was sent up from Omagh the day before to join the two leading battalions whose task was to break into the Bogside and Creggan estates in Londonderry and re-establish the security forces in what had become an IRA 'no-go' area. The preliminary briefing was hardly encouraging, with intelligence of booby-trapped barricades, anti-tank mines, machine-gun positions etc. In the event the worst did not occur. The armoured cars were into both estates within minutes of crossing the start line, skirting round the barricades and smashing their way through a succession of garden fences to reach their objectives. Opposition was light because the IRA had realized the game was up and had withdrawn to Donegal in order to fight another day. The quick success was in some ways an anti-climax but nothing could be taken for granted. During the first night the squadron second-in-command ventured out of his armoured car only to be greeted by a sharp burst of gunfire; he withdrew into his vehicle rapidly.

Operation Motorman, July 1972

The Squadron had been allocated St Joseph's RC School for
its billets and took up residence in the school gymnasium. As a
consequence the Squadron Leader was able to send his first
dispatch to Omagh on the Headmaster's headed writing paper.
The Creggan remained quiet but the Bogside continued to stir
up trouble. There were also regular sniping attacks on the Army
and the RUC. In the Bogside armoured cars were greeted with
paving stones, ripped up from the streets below and thrown
down from upper windows and roofs. 10-year-old children
achieved unbelievable feats of accuracy in throwing sharp
stones and pieces of slates into any aperture or hatch carelessly
left open in the armoured cars. It was quickly realized that
patrolling by armoured cars could be provocative and soon the

Infantry took over. After a week there was little left for A Squadron to do and it returned to Omagh.

At the outset in Northern Ireland the lessons of earlier internal security campaigns, particularly those relating to Intelligence and the Joint Management of Security resources, were not fully implemented. As always 'cost' was the yardstick applied by the financiers, regardless of the fact that essential measures deferred today would become even more costly, both in blood and treasure, when implemented in the future. One example of this mistaken approach was the construction of the base for the Army in County Fermanagh. Lieutenant-Colonel Dennis was convinced he had won the argument for it to be sited in the new UDR barracks which were being built in Enniskillen, thus co-locating two of the principal elements of the Security Forces. 'Imagine my surprise — indeed anger,' he writes, 'when from my helicopter I saw building proceeding on the old airfield at St Angelo some 10 miles from the expected site. This ensured that the undesirable split between Army, UDR and RUC was perpetuated. The site had other serious drawbacks but all these arguments were set aside in favour of reducing costs.' It was ever thus!

The Regiment's services in Northern Ireland were well recognized in the list of awards. In addition to the OBE for Lieutenant-Colonel Dennis there was one MC (Major Radford), one MM (Sergeant Causer), three George Medals, two MBEs, three BEMs, seven Mentions in Despatches and five GOC's Commendations. But there was a price to pay. Corporal Powell was killed by a land-mine on 28 October, 1971, and 2nd Lieutenant Somervell on 27 March, 1973, when on a routine patrol only 10 miles from Omagh. Lieutenant Clive's miraculous escape has already been described. Within the Regiment's area of operations eighteen other members of the Security Forces were killed (nine Army, seven UDR and two RUC) and forty-three were wounded. The War Diary records 134 shooting incidents, 207 explosions, and 134 bomb hoaxes. Twenty-six weapons, including one RPG 7, 12,274 pounds of explosives, and 1,974 rounds of ammunition were recovered.

Of all the British Army's internal security campaigns,

Major John Wright discussing operations with an officer of the Ulster Defence Regiment and a constable of the Royal Ulster Constabulary

Northern Ireland is probably the one most in the public eye, due to its closeness to home and the influence of the media. During its stay in Omagh the Regiment became extremely publicity conscious for probably the first time in history.* The monthly newsletter *The Lancer* was quoted several times in the national newspapers, and a stream of journalists, both national and from Staffordshire, visited the Regiment. Many interviews were given to local radio and press, and on three occasions on National television. Whether any correlation can be made between this publicity and recruiting is difficult to determine, but the fact remains that having arrived in Omagh well under strength, by the time it came to leave the 16th/5th Lancers was the best recruited regiment in the Royal Armoured Corps. It is also worth comment that when volunteers were called for to join the 1st Royal Tanks who were taking over in Omagh, no less than 150 soldiers of all ranks said they were willing to forgo Cyprus and Hong Kong to continue to serve in Fermanagh and Tyrone — a clear indication of the sense of purpose all ranks had while in Northern Ireland. The tour there ended on 23 May, 1973.

* It is probably safe to assume that towards the end of the last century the 16th Lancers would have considered such publicity to be 'extremely bad form'! But times have changed and WO I Hopkins handled the Regiment's Public Relations in Omagh with great skill.

CHAPTER THIRTEEN

Cyprus & Hong Kong
1973-74

THE Regiment returned from Omagh to Tidworth in May, 1973. Regimental Headquarters, Headquarters and the Air Squadrons were to be based there. A Squadron donned blue berets and set off for Cyprus as part of the UN Peacekeeping Force (UNFICYP); they left their families behind in Tidworth. B Squadron also left for Cyprus but as part of the permanent British garrison in the Sovereign Base Area (SBA) in Dhekelia; they took their families with them. As did C Squadron on an 18-month tour in Hong Kong as part of 48th Gurkha Brigade. A and C squadrons were to be equipped with the Scorpion light

The Regiment drive through Kowloon on the Queen's Birthday Parade

tank, but B Squadron still had Saladins, Saracens and Ferret scout cars. It was a strange amalgam of armoured fighting vehicles.

In January, 1974, Lieutenant-Colonel Morris succeeded Lieutenant-Colonel Dennis as Commanding Officer. He had been Second-in-Command for most of the time in Omagh. Originally in the 3rd Hussars, he had spent some time with the Federation Armoured Car Regiment in Malaya when it was virtually a 16th/5th Lancers preserve under Lieutenant-Colonel Keith Robinson. Morris joined the 16th/5th Lancers on the amalgamation of the 3rd Hussars with the 7th Hussars. Another who joined the Regiment at the same time, also from the Malayan Armoured Car Regiment, was Major Gauntlett, formerly from the Royal Tank Regiment.

When Cyprus became independent in August, 1960, Britain retained two Sovereign Base Areas at Dhekelia and Akrotiri. In 1963-64 there was a civil war between Greek and Turkish Cypriots that was ended only by UN intervention. Turkish Cypriots established their own enclave which among other things divided Nicosia by the 'Green Line'. The UN set up a peacekeeping force to keep the two warring communities apart. Intended only as a temporary measure, UNFICYP is still in existence 25 years later. Turkish and Greek Cypriots continued to quarrel. On 15 July, 1974, there was a coup against the Cypriot President, Makarios, who managed to escape to Britain via Malta. The Turks sought British aid to restore the republic but Britain drew back. The Turks therefore decided to go it alone and started to invade Cyprus on 20 July, 1974. Twenty-four hours earlier, at 0430 on 19 July, RHQ, HQ and A Squadrons were placed at 24 hours' notice to be flown out to Cyprus; this notice was reduced that afternoon to 12 hours when Lieutenant-Colonel Morris was told he would be required on arrival in Cyprus to command a composite regiment consisting of B Squadron Blues and Royals; C Squadron 4th/7th Royal Dragoon Guards, and his own A and B Squadrons, the latter on the Regiment's arrival in Cyprus.

Air Marshal Sir John Aiken was Commander British Forces Near East. The GOC Near East Land Forces was Major-General

Observing the
Turkish front
line from a
Saladin
armoured car

H. D. G. Butler. The Commander UNFICYP was Lieutenant-General Prem Chand of the Indian Army, whose Chief of Staff was Brigadier Frank Henn of the Royal Hussars. There were plenty of 'Chiefs' for not all that many 'Indians '! When trouble broke out Aiken's main concern was for the British families, 4,500 of whom were living outside the Sovereign Base Areas in private accommodation. They were hostages to fortune in what was predominantly Greek Cypriot territory. Were it to be assumed that HMG was looking on benignly while Turkey invaded the island, these families would be at risk, as also would be the Sovereign Base Areas. If, on the other hand, British troops became involved in hostilities with the Turks, who could reinforce their troops very quickly from the Turkish mainland, we looked like having the worst of things. It lent force to the arguments of those who had from the very beginning insisted that the Sovereign Base Areas would become hostages to fortune in the event of war between Greece and Turkey, and that therefore the best thing we could do about Cyprus was to leave it lock, stock and barrel.

Turkish troops began to land at Kyrenia on 20 July; within 48 hours they had established a 15-mile wide corridor between Kyrenia and Nicosia. The Greek Cypriot National Guard, supported by Greek troops flown in from the mainland, resisted strongly, but the heavier armoured and air support of the Turks began to tell. The Turks bombed Nicosia airfield, the Greek quarter in Famagusta, and even succeeded in sinking one of their own frigates. A ceasefire was then somehow arranged, although no one expected it to last. It did, however, permit the numerous families living outside the Sovereign Base Areas to be gathered in to safety.

B Squadron, operating from Pergamos Camp on 20 July, were heavily involved. '1,000 cars and lorries carrying 4,500 British and foreign civilians were escorted from the Nicosia battle zone to the safety of the British base at Dhekelia. The convoy, flying Union Jacks, was guarded on the 35 mile journey by armoured cars of the 16th/5th Lancers ' [*Daily Telegraph*, 22 July 1974]. 3rd Troop was tasked with getting the staff out of the British High Commission but the gates had been securely locked and the keys could not be found. In no way disconcerted, Lieutenant Orr-Ewing burst through them with his Saladin to get everyone out. 'The next day, after repeated heavy bombing on Famagusta, in went the Lancers squadron again. Under intense air and ground attack (accompanying Royal Marines were wounded and a small child was killed), they got 1,500 people out to Dhekelia ' [*Daily Telegraph*].

The air move from Tidworth to Cyprus left much to be desired. The RAF, having been told to get everyone to Cyprus in the shortest possible time, paid no attention to what might be described as tactical loading. One consequence of this was that the binned 4 tonners carrying spares for the LAD were scheduled to arrive before the Commanding Officer and his RHQ vehicles. Luckily the LAD vehicles were too high to fit into the aircraft when they tried to load them. There was further confusion on arrival at Akrotiri to discover that HQ Cyprus had decided, without informing the UK or the troops involved, that there was not to be a composite regiment after all. The Blues & Royals and 4/7 DG squadrons were to remain at Akrotiri under command of 19 Brigade. Lieutenant-Colonel Morris and A Squadron were to move to Dhekelia, where B Squadron was already installed, leaving behind HQ Squadron with the LAD, and the Echelon with spares, fuel and ammunition. Furthermore, Lieutenant-Colonel Morris was not to command anything, composite or non-composite, but was instead to function as GSO I to the Commander Dhekelia SBA. It took him some 36 hours to unscramble this nonsense, after which he was permitted to return to Pergamos Camp to command what there was there of the Regiment.

Everything changed at 2015 on 24 July when the Regiment was put at 12 hours' notice to move to Nicosia. Not A Squadron as expected, which had been ordered for political reasons to leave their 17½-year-old soldiers behind in Tidworth, but B Squadron who, in a normal peacetime garrison, had their under-18s with them! The reason for this switch was because A Squadron had Scorpions which, being tracked, would create political ructions if employed outside the Sovereign Base Areas, whereas B Squadron's vehicles were all wheeled. A Squadron could not therefore be employed in an area as sensitive as Nicosia.

At 2230 RHQ left Pergamos Camp ahead of B Squadron. C Squadron 4/7 DG, HQ Squadron 16/5L, and a company of the Coldstream Guards were to follow the next morning at 0800, plus an RAF officer as Forward Air Controller, if they could get through. The 16th/5th Lancers, now an embryo 'Reconn-aissance Battlegroup', had been given Phantoms as air support, gained some Foot Guards and detached the Blues & Royals, while retaining 4/7 DG with their powerful Swing-Fire missile troop.

There is a good description of B Squadron's move to Nicosia to join the UN force there in the 1974/75 edition of *Scarlet & Green*:

'The Commanding Officer went on ahead* of the column which arrived in what appeared to be an almost deserted city at 0200 hrs. Little was known of the situation in Nicosia at that time and it was an uncomfortable and slightly sinister feeling driving through the empty streets without knowing quite what to expect. RHQ was guided into Gleneagles Camp and B Squadron formed up on the old Airport road to await further instructions. Gleneagles Camp did nothing to restore our morale or confidence; it was pitch dark and derelict-looking and

* To attend a briefing by Brigadier Henn, Chief of Staff, UNFICYP.

there was no one there to meet us. But shortly after our arrival a Land-Rover appeared bearing a pile of blue berets, brassards and all the other paraphernalia and in the dark we struggled to transform ourselves into United Nations troops.

'No doubt we looked more like Fred Karno's Army but in the height of the confusion the Commanding Officer appeared and called for instant orders. What he had to say was short and to the point but it was of such a nature that it would be wrong not to record it in its entirety. He said, "I have no time to waste. I have just seen the Brigadier and he expects the Turks will attack the airport at dawn – in a few hours' time. They are at battalion strength and have a company of M 47 tanks and artillery in support. We are to hold the airport with B Squadron and you [RHQ] are to take up positions here. I am going off to brief Major Wright ['B' Squadron leader] now. Any questions"

'After a moment of stunned silence the Second-in-Command [Major Harry Gauntlett] tentatively enquired as to which direction we might expect the enemy and the Colonel replied in the grand traditional manner. "There is the enemy," he cried, flinging out his arm and pointing into the inky blackness and then hastily departed. No time was wasted in preparing our defences and by dawn we were ready. But as the first light began to break over the Kyrenian Hills behind us we knew instinctively that something was wrong, very badly wrong. We were facing in completely the wrong direction. Undaunted, we turned about and prepared to meet the assault once more but nothing happened and after an hour or so we stood down and relaxed.'

This light-hearted description of what must have been a very tense situation at the time is in the best traditions of the British Cavalry which always has found it best not to take war too seriously.

B Squadron's situation at Nicosia airport as the sun rose on 25 July was distinctly nasty, but morale was high. Mention has already been made of the part played by the Squadron in the evacuation of British and foreign families from Nicosia and Famagusta between 20 and 24 July. This difficult task had been so successfully accomplished that General Butler had issued the

Observing the
Turkish positions

following Press Release: 'The 16th/5th Lancers have added to
their distinguished reputation in the last few days in Cyprus,
when, faced with the task of escorting the evacuation from
Nicosia and Famagusta, they played a vital part in enabling the
concentration in the Sovereign Base Areas of virtually every
British subject and many people of other nationalities.'
Moreover, just prior to the outbreak in Cyprus, the Squadron
had been working overtime in readiness for an inspection by
the GOC intended to establish whether the Squadron was 'Fit
for Role '. General Butler was, however, so impressed with the
Squadron's performance between 20 and 24 July that, just prior
to the move from Pergamos Camp to Nicosia, he formally
agreed that no further inspection was required. This provided
a tremendous boost to morale.

The confirmatory signal from the GOC was received while
the Squadron was preparing for action early on 25 July when
it seemed that his confidence would very shortly be put to the
test. Lieutenant-Colonel Morris had been ordered to hold the
airport for the United Nations. Most unusually for what was a
United Nations operation, he was specifically ordered to open
fire should this be necessary. At the same time the Battle Group
was to reconnoitre the Turkish positions west of the
Nicosia-Kyrenia road. Perhaps this was the first time (certainly

since Korea) that a British CO was empowered to start a war at his own discretion! His troops consisted of B Squadron, with Saladins and Ferrets; 30 Finns; 200-plus Canadian logistic troops; and 10 Swedes in two remarkable Scania-Vorbis TLV armoured personnel carriers.

Opposing this scratch force was a Turkish battalion group, which included a number of M47 and M48 tanks – medium tanks which out-gunned and out-armoured the Saladin armoured cars. They were approximately 1,000 metres from the airfield. Were they to advance, it would be impossible to avoid a 'heavy metal firefight if they breached the perimeter 'No-Go' line 300 metres in front of our forward defensive position. Fortunately they did not. As Intelligence sources later confirmed, having heard the menacing howl of the Saladin engines as they moved into hull-down positions before first light, as well as the burst of cheering from the Canadian logistic troops when they saw us arrive, the Turks were anxious to avoid involvement with a UN force which they knew now included a very formidable unknown quantity – British manned AFVs! '

There were some tense hours of 'eyeball to eyeball' confrontation before the situation began to cool down. Then the Finns, Canadians and Swedes departed, to be replaced by C Squadron 4/7 DG and a company from the Coldstream Guards. Meanwhile there was fierce fighting between the Turks and the Cypriot National Guard, substantially reinforced by Greek regulars, as the Turks tried to widen the corridor to Kyrenia, where they were continuously landing troops. There was a great deal of confusion regarding the precise location of the 'Front Line ', confusion that was repeated on the international level. The Turks facing the Lancers from below the airport insisted that it was in their hands. General Prem Chand maintained that the UN force had taken it over and that therefore it was temporarily internationally protected. Britain moved eight RAF Phantoms to Akrotiri on 25 July, although no one wanted to get involved in war with Turkey, a NATO and CENTO ally. Fortunately the extreme right-wing government in Athens collapsed and the new Prime Minister, Karamanlis, was much more reasonable. Likewise in Cyprus Sampson was replaced by Clerides, and it became possible to negotiate a ceasefire agreement which came into effect on 30 July. Service families were then allowed to return to their homes outside the Sovereign Bases.

It was a very uneasy truce. Having established more or less the front lines, B Squadron found itself umpiring between the two sides, 'neither of which trusted either us or each other '. The Turkish build-up through Kyrenia continued and there were continual clashes between them and the National Guard, fortunately without causing any casualties in B Squadron, although there were some near misses. On 14 August, however, the Turks, having demanded that roughly one-third of the island should become a Turkish Cypriot zone, launched a full-scale assault. It began with artillery fire and air strikes, in the course of which 6th Troop of B Squadron (2nd Lieutenant Goodwin), though in white-painted scout cars, in the open, and flying the UN flag, came under air attack by Turkish Super-Sabres at Ayia Marina where they had an Observation Post. Four men were wounded, three of them severely, but fortunately all were successfully evacuated to Gleneagles Camp. Lance-Corporal Wood, who acted with great courage, later received the Queen's Commendation for Gallantry.

Caught once again with their families largely located outside the Sovereign Bases, the British had to mount a hasty operation to bring them back into the fold. This task fell to A Squadron with its Scorpions from Pergamos Camp, while in the north, at Nicosia, B Squadron sat on the airfield. There were numerous incidents on the perimeter of the Dhekelia SBA, including the one of the Turkish M 47 tank which ran out of petrol close to a roadblock manned by 3rd Troop. It had also run out of main armament ammunition, the secondary armament was jammed, the radio did not work, and the tank commander had no map. In Lieutenant-Colonel Morris's opinion the Turks were very slow, both in planning and execution: 'Only during the advance to Famagusta did the Turkish mechanized groups really motor — and then a whole Division was I think lost. ' [its way, not the battle!]

The High Commissioner's
Daimler

The story of the High Commissioner's Daimler is worthy of mention. The British High Commission in Nicosia found itself in the front line when the Turks attacked on 14 August. It soon became urgently necessary to evacuate the High Commissioner, his wife and the staff. Following a request by the High Commissioner and a rapid change of plan, a detour was made to extract the Italian Ambassador and his wife. This was carried out initially by the Commanding Officer and the Adjutant in the Command Saracens with 7th Troop, who continued, once clear of Nicosia, as escort to the High Commissioner to Dhekelia. It was during this episode that the Commanding Officer had to explain a somewhat garbled wireless conversation: 'SUNRAY is heading south locked in the back of a Saracen with an Italian lady and he will be unable to speak for 30 minutes'!

Before leaving the High Commissioner, the Commanding Officer asked him for the keys of the official Daimler which had been left behind when leaving Nicosia under fire. It was successfully recovered and according to one SITREP 'the effect this vehicle has on people is amazing! The Canadians thought it was Dr Waldheim [UN Secretary-General] when the CO drove up, and its effect on the Turks has to be seen to be believed!' As, for example, on 17 August when 'The Turks devoted their time to consolidating their line and in the area of our front they pushed towards our perimeter. The CO immediately drove off to see them in the borrowed High Commissioner's Daimler and in the most imperious tones demanded their immediate withdrawal. Rather surprised by the turn of events, a whole company of Turks rose to their feet and meekly complied. ' Later the Second-in-Command, under a battery of Turkish TV cameras, was berated by the Turkish military commander and political commissar and subjected to a half-hour discourse on the British Imperial system. Sadly, the Daimler had soon to be returned to its rightful owner.

Eventually a precarious ceasefire was patched up between the two contestants. Despite numerous shooting incidents, it managed to hold, chiefly because the Turks were in such overwhelming strength that nothing short of all-out war between Greece and Turkey could be expected to dislodge them, and probably not even then. Cyprus was now divided, the north a Turkish enclave, with 150,000 of its former Greek Cypriot inhabitants now refugees in their own country. Although a wholly unnecessary war, it had provided the Regiment with valuable combat experience and had won for it an excellent reputation in both British and UN circles. It had gained no fewer than thirteen awards for distinguished service or gallantry, including one Queen's Gallantry Medal. Additionally, and most unusually, by special directive from UN Headquarters in New York, every officer and soldier who served in UNFICYP as part of the '16th/5th Lancers Reconnaissance Battle Group Kyrenia West ' during the short Cyprus war, was awarded the UN medal, usually only awarded after completion of 90 days' service with a UN force.

Once the ceasefire held, and before leaving Nicosia, the Regiment held a formal parade at the derelict airport, inviting General Prem Chand to take the salute. He wrote later 'to express our sincere gratitude to you and all ranks under your command for the splendid manner in which the Regiment has carried out its duties and responsibilities during the recent operations in Cyprus... May I also take the opportunity of thanking you for the privilege and honour of taking [the salute at] your parade the day before you left Nicosia. This will always

Farewell Parade at Nicosia Airport. The Drive-Past the UN Commander General Prem Chand

remain a memorable occasion for me, especially in view of the time-honoured associations the Indian Cavalry has had with the British Cavalry dating back to the days of the Bengal Lancers and beyond. '

The Regiment also received a mention in the House of Commons, an unusual honour, when the Foreign Secretary, James Callaghan, told the House on 31 July: 'The 16th/5th Lancers and others have played a very steadying role in what could have been a most critical situation.'

It had been one of the most unusual, and most complicated, episodes in all the Regiment's long history. Peacekeeping produces no Battle Honours but it is probably more in keeping with the end of the twentieth century than many a bloody battle of the past. It requires patience, resolution and discipline of the highest order. We can be proud that when the Regiment was put to the test it came through with flying colours.

On relief by the Queen's Royal Irish Hussars, B Squadron returned to Pergamos Camp for the duration of its overseas tour. A Squadron flew back to England. RHQ and HQ Squadron went to share Paphos Camp with the Coldstream Guards until they flew home to Tidworth on 15 September, 1974.

CHAPTER FOURTEEN

Beirut 1983-84

IT may seem strange in a history covering 300 years to devote a complete chapter to A Squadron's three months' stay in Beirut over Christmas, 1983. However, the Squadron's task was so unusual, calling for qualities of patience and restraint not normally associated with active service, that it seems to this author to be worthy of separate record. But first it has to be explained why British troops came to be deployed in Beirut.

The tragedy of the Lebanon first began to unfold in 1975 when Christian and Moslem Lebanese came to blows. It has continued with brief ceasefires to this day, until Beirut, once the most civilized city in the Middle East, has been reduced to ruin. The ramifications of Lebanese politics are too complicated to be explained here but in essence it is a religious conflict, which has been further complicated by the Palestinians (PLO), who took refuge in Lebanon in 1971 after King Hussein had driven them out of Jordan. The PLO, in its turn, was driven out of Beirut in 1982 by the Israelis. By then Syria had entered the conflict, ostensibly in support of its Moslem co-religionists, but actually in an attempt to establish its authority in a country which had formed part of the province of Syria under the Turks. Syrian troops in large numbers have been deployed in Lebanon for the past ten years but their presence has done nothing to bring peace to that wretched country.

Like all civil wars, the struggle is waged with total disregard for the civilian population. No rules govern the conduct of those engaged in the fighting. Horror after horror has been perpetrated by both sides. No quarter is asked; none is given. There have been massacres, car bombs, kidnappings, shelling of built-up areas, skulduggery of every kind, assassinations, betrayals and open gang warfare. In April, 1983, after the Israelis had pulled back from Beirut, 241 US Marines lost their lives in a bomb attack on their barracks; they were part of the multinational force sent to oversee the Israeli and PLO withdrawal. The French contingent also suffered severely. And yet, despite this mayhem, some Lebanese still contrived to make fortunes. It is typical of the paradox which is Lebanon.

When it was first proposed to establish a multinational force in Beirut in order to monitor the PLO's withdrawal, followed by the Israelis', it was an open secret that the British chiefs of staff were unenthusiastic. The government was no keener. The British had burnt their fingers so often and so disastrously in

Patrolling in downtown Beirut

the Middle East that they preferred to let the Arabs go on murdering each other rather than intervene to prevent them from doing so. Eventually, however, as a result of American pressure, HMG reluctantly agreed to provide an armoured reconnaissance squadron as part of the multinational force. The first regiment to provide it was the Queen's Dragoon Guards. The Regiment was warned that it would relieve the QDG in June, 1983, with a squadron equipped with Ferret scout cars, but in view of the forthcoming Guidon Parade, and with the very sporting agreement of the QDG, the relief was postponed until later in the year. Lieutenant-Colonel John Wright was in Denmark on a CPX in September, 1983, when he was told that the QDG squadron was to be relieved in December. He accordingly selected A Squadron (Major Robin Faulkner) for the task, and training for it began in October. They arrived in Beirut in early December.

There was in fact no precedent for the kind of training required. After Northern Ireland the 16th/5th Lancers were experienced in dealing with both urban and rural terrorism; but nothing in Belfast nor County Armagh resembled conditions in Beirut where men went about armed to the teeth and needed no encouragement to fire their weapons whenever the fancy took them, regardless of the crowds going about their business in the streets. Nor was the firing restricted to small arms. Artillery fire was normal, as was mortar fire. Explosives, grenades and rocket launchers formed part of the armoury. Unless one was a Lebanese, and belonged to one of the many private armies, it was virtually impossible to distinguish friend from foe. Inability to speak Arabic merely compounded the difficulties.

One of the Squadron's principal tasks was to supervise the meetings of the so-called 'cease-fire committee' on which all the chief protagonists were represented; they were representatives of the Druse, Shi'ite Moslem militias, Christian militias, and the Lebanese Army. These meetings took place in a bombed-out bank in the No-Man's-Land between West (Moslem) and East (Christian) Beirut. At one time they seemed to be the only hope for a negotiated settlement but like everything else in the

Lebanon, 'Hope is a good breakfast, but it is a bad supper' . They came to nothing in the end. It was the Squadron's task to provide security for the cease-fire committee when it met, supposedly daily but in fact at very irregular intervals. After searching the surroundings, as well as the rubble in the shell of a bank, and deploying six scout cars round the building for protection of the delegates, it was a matter of waiting for the delegates to turn up. Sometimes they did; sometimes they didn't. 'I got to know the delegates well, ' says Faulkner, 'often taking their sweet and very strong coffee with them. They all treated each other like long-lost friends but then returned to their own lines and often started shelling.'

He goes on to say that 'There were some very tense moments. The guarding was less of a military operation but more an act of faith.' On one occasion the Shi'ites (AMAL) claimed that the Squadron as deployed round the bank was covering an advance by the Christians (Falange). Unless the Squadron withdrew at once, it would be wiped out. Clearly to do so without any apparent reason would destroy the Squadron's credibility as a security force. Faulkner went down to the bank himself to assess the situation. 'It was a lonely ten minutes' drive down,' he says, 'deliberately slow, but as normal as possible so as not to alarm the unseen watching gunmen. The air was unnaturally still, even the insects made a deafening sound as I walked round, visiting each scout car in turn. We were very relieved when after two hours we were told that there was to be no meeting and we withdrew as unhurriedly as we had driven down – this time a little more thankful when we got back to base.

'The Squadron was also called upon to send out patrols. West Beirut is a maze of streets which were invariably crowded whenever the shelling and firing had died down. It was easy to get lost. There was also the uneasy feeling, when standing up in the turret of a Ferret scout car, that one provided an uncommonly good target for anyone with a Kalashnikov handy. Somehow or other we managed to get away with it. In fact there was probably more danger from the notorious driving of the Lebanese than from the trigger-happy gunmen.'

'We have had to adapt quickly to the Beirut style of driving

Major Faulkner (centre) and Captain Ghillyer survey their area of responsibility in Beirut. *Terry Fincher*

and we give as good as we get,' said Sergeant Leaning. 'It had meant the odd knock here and there but we fancy that with four tons of Ferret around us, the odds are on our side.... We mainly go out to establish our presence. But we are always on the alert and are very much aware that someone might suddenly decide to have a go at us. Overall we get a friendly reception. We are here as peacekeepers, not to take sides.... I think the Lebanese appreciate that.'

Although every man in the Squadron was well aware that they were operating in an environment where the normal scale of human values did not apply, there was no intention of adopting a kind of troglodyte existence. From the outset Major Faulkner insisted that they lived in as civilized a fashion as conditions permitted. The former cigarette company's offices that were the base of the British contingent could hardly be described as a palace but it did possess a splendid flat roof which provided a fine vantage point for surveying the mayhem going on below. It was of course vulnerable to both shell and sniper fire but this did not deter the numerous press correspondents from coming up to have a look, not infrequently glass in hand, since A Squadron dispensed hospitality as it should be dispensed in an Arab country.

In order to create one team of the various elements that made up the British contingent, Major Faulkner established unified Officers' and WOs' & Sergeants' Messes. They even contrived to find a dining table large enough to seat the officers, and embellished it with a 16th Lancers silver horse as centrepiece. They entertained a US Marines Forward Observation Team, as well as liaison officers from the Lebanese Army. Relations with the press were extremely good, and support from the Royal Navy and Royal Air Force was first-class, particularly the helicopter pilots, Chinooks and Sea Kings.

'Overall, life was a hard routine grind,' says Major Faulkner, 'maintaining a balance between keeping people safe, keeping morale high and achieving the aim of being there, under very trying, overcrowded and often dangerous conditions.... There were some lighter moments, as when an Arab Stud was found which had just had a miraculous escape from a ground-to-ground rocket which had landed in the exercise yard... or when some of us went skiing and the ski lifts were guarded by the Falange... or when being taken out to a Lebanese restaurant by the Lebanese Liaison Officer!' There was of course always the risk of a stray shell or mortar bomb, not to mention the kind of suicide bomber who wreaked such havoc in the case of the US Marines. 'Our operational stance reflected our political role – we had to look efficient and professional, to *deter* attack, yet not aggressive; that might *invite* attack.' A delicate balance.

In addition to playing host to a great many correspondents, A Squadron found itself entertaining even more distinguished guests who included the Secretary of State for Defence, Michael Heseltine, and the Chief of Defence Staff, Field-Marshal Lord Bramall. All came away remarking on the high morale of every Lancer they met, as well as commenting on the very professional way they went about their duties. But as time went on the high hopes of an agreement being reached between the various Lebanese warring factions, with which A Squadron's mission had started, began after Christmas to vanish. January, 1984, saw a rapid deterioration in the situation; the streets, dangerous before, became doubly so, and the shelling and machine-gun fire grew in intensity and volume. There was a gradual breakdown in the reconciliation process and the Security Committee, for whose safety the Squadron was responsible, met for the last time on 16 January, 1984. None of the parties concerned evinced any intention of coming together again.

From late January onwards heavy fighting made street patrolling both irrelevant and dangerous. It also increased the likelihood of the Force's base being hit during the exchanges of fire. Supplies were running short and the mains water supply was cut on 5 February. There was little or nothing A Squadron could do to remedy matters and on 7 February the decision was suddenly made to remove the British contribution to the Multinational Force. This came as a complete surprise (the Squadron Leader was away on 4 days R & R leave in Cyprus at the time) but the withdrawal operation was most capably handled by his Second-in-Command, Captain Julian Snell.

It entailed backloading all the vehicles and kit to the port of Jounieh (in Christian East Beirut), whence they were transported by Sea King and Chinook helicopters to the RFA *Reliant* lying offshore. It involved no less than 80 sorties but by just after last light on 8 February the operation had been completed. A Squadron had left Beirut, as it had arrived, by courtesy of the Royal Navy pilots of the Sea Kings and the Royal Air Force pilots of the Chinooks, to whom Major Faulkner pays tribute when he writes: 'I can honestly say that without the leadership and the determination of the Chinook and Sea King detachments, and the Captain of the RFA *Reliant*, our speedy and dramatic withdrawal could have ended in disaster.'

The Colonel-in-Chief and A Squadron meet by chance in Cyprus

Since peacekeeping is likely to become ever more prominent in the peacetime tasks of the British Army, A Squadron's experience in Beirut may serve as an excellent example of how best to deal with what will always be a dangerous and very confusing task, demanding discipline of the highest order and, additionally, great patience and not a little diplomacy. Certainly A Squadron displayed all these qualities as was made clear in a letter to the Commanding Officer from General Sir Frank Kitson, C-in-C United Kingdom Land Forces, on 28 March, 1984: 'Your soldiers' performance was a credit to the Regiment and to the Army as a whole. They earned and deserved the respect and confidence of the other members of the Multinational Force. This was achieved in circumstances which were always difficult and often dangerous.'

It was a well-deserved tribute. Major Faulkner, 2nd Lieutenant Bullen, Staff Sergeant Lock, REME, and Sergeant McDermott, ACC, were Mentioned in Despatches, and Corporal Trower and Trooper Hawkins received the GOC's Commendation. To the Squadron's huge delight, however, the Beirut operation was rounded off in spectacular fashion when the Queen, en route to Jordan on a State Visit, stopped over briefly at Akrotiri in Cyprus and asked to see A Squadron of her Regiment. 'I was proud to present them to Her Majesty,' writes Major Faulkner, 'the trumpeter having sounded and with lance pennants fluttering, drawn up in front of the royal aircraft.' It was a fitting end to what had been one of the most unusual episodes in the long history of the Regiment. On 1 April the Squadron returned to Tidworth.

CHAPTER FIFTEEN

The Gulf War
15 January - 28 February, 1991

IRAQ invaded the independent Arab state of Kuwait on 2 August, 1990. Although Iraq had laid claim to Kuwait almost since its own inception as an Arab state in 1920, the ostensible cause for the invasion was Kuwait's refusal to cancel the debt incurred by Iraq as a result of the financial assistance it had received from Kuwait during Iraq's war with Iran from 1980-88. The Iraqi dictator, Saddam Hussein, had built up a large and well-equipped Army and Air Force which quickly overwhelmed the Kuwaitis. The Emir of Kuwait and the rest of the ruling Al Sabah family fled to Saudi Arabia as the Iraqis poured into Kuwait. Ten days later Saddam Hussein annexed Kuwait as Iraq's nineteenth province.

The reaction of the United Nations to this act of bare-faced aggression was surprisingly swift. Under American leadership the Security Council passed Resolution 660 on the very same day as the invasion by 14 votes to nil, with Yemen abstaining. This condemned Iraq's invasion of Kuwait and demanded an unconditional withdrawal, this to be followed by negotiations between Iraq and Kuwait. It was to be the first of twelve Security Council Resolutions that were passed during the Gulf crisis, the most important of which were Resolution 661, which imposed a trade and financial embargo against Iraq, and Resolution 678 (on 29 November), which authorized member states to use all necessary means to make Iraq withdraw from Kuwait if Iraq had not done so by 15 January, 1991.

An international coalition was formed under American auspices to bring political and military pressure to bear on Iraq, but Saddam Hussein appeared to be as immune to military threats as he was to diplomatic persuasion. Britain's land contribution to the coalition force was two Armoured Brigades grouped to form the 1st (UK) Armoured Division. The Divisional Commander was Major-General Rupert Smith, and the overall Commander of the British Land, Sea and Air Forces was Lieutenant-General Sir Peter de la Billière. The Commander-in-Chief of the Allied Coalition Forces was General Norman Schwarzkopf of the US Army. The British armoured regiments were equipped with the Challenger Mark I Main Battle Tank, as yet untried in battle, as indeed were virtually all the British officers and soldiers involved.

As the slide to war gathered pace during the last few months of 1990, the Regiment was preparing to move from BAOR to the United Kingdom in relief of the 9th/12th Royal Lancers at Wimbish in Essex. This was due to take place in January, 1991. As an armoured reconnaissance regiment it was equipped with a series of vehicles called Combat Vehicle Reconnaissance (Tracked), or CVR(T). These were not in their first youth. They weigh about 10 tons and have a short and narrow track when compared with a main battle tank like Challenger. This limits their speed of movement across country, as was found to be the case in the Saudi and Iraqi desert. They are able to operate closed down, have NBC protection, and can operate at night by using night vision equipment. Communication is by radio.

The Regiment had several variants of the CVR(T), the principal one being the Scimitar. This has a 30mm cannon for protection and a crew of three — commander, gunner and driver. The anti-tank guided weapon vehicle is the Striker, which also has a crew of three. The Spartan is a personnel carrier of up to five or six men, including a commander and driver; it also carries equipment for mine-clearance, anti-tank and infantry weapons, demolition equipment, etc. Other vehicles in the series are the Sultan command vehicle, packed with communications equipment; the Samson for recovery; and the Samaritan ambulance. The CVR(T) is only lightly armoured and does not provide anything like the same degree of protection as a main battle tank. The Regiment also had a range of B vehicles which included Land Rovers and trucks for the carriage of ammunition, fuel, rations and water.

With the impending move back to England, everyone was busily engaged in preparing the vehicles and barracks for handover to the incoming regiment. Also all the Regiment's B vehicles had been transferred to 7th Armoured Brigade to bring them up to war establishment. Moreover, a large quantity of other equipment and spares had been handed over to 7th Armoured Brigade, who by now were in the Gulf.

It was towards the middle of November that the Commanding Officer, Lieutenant-Colonel Philip Scott, was due to dine with the Commander Armoured at HQ 1st (British) Corps at Bielefeld. He was asked to arrive 15 minutes early and it was then he was told that a request had been made for a complete armoured reconnaissance regiment to be sent to the Gulf; at the time there was only A Squadron of the 1st The Queen's Dragoon Guards (QDG) providing reconnaissance for 7th Armoured Brigade in the Gulf. Were the request to be agreed, the 16th/5th Lancers would be the regiment chosen. This was confirmed on 22 November, and from then onwards it became a frantic race against time to prepare to move to the Gulf rather than to the United Kingdom. Exactly a fortnight later the Regiment's vehicles were loaded on to the ships for the Gulf. During these two weeks the Regiment had to take in a complete 'new' B vehicle fleet plus a considerable influx of additional A and B vehicles. Many of these vehicles were in an appalling state. In addition the whole vehicle fleet had to be painted in desert camouflage. All the time manpower reinforcements were arriving to bring the Regiment up to war establishment. The Scimitars and Strikers with crews all went up to Hohne Ranges to fire. Every man underwent a package of individual training to help prepare for desert conditions and the Regiment did an exercise at the Battle Group Trainer. The equipment and stores flooding into the Regiment made for a Quartermaster's nightmare.

As fate would have it, most of the British Army's stocks of

desert combat suits had been sold to the Iraqis some years previously. A new issue was being frantically manufactured and was to be issued on arrival in the Gulf. Every man was sized before leaving Germany, but not always to useful purpose apparently. Corporal Atkinson's trousers were two feet too long, we are told. It cannot be said that the new style of combat dress lived up to the high standard of turnout of the British Army when it had previously gone to war. However, it was certainly more practical.

The Regiment began arriving in Saudi Arabia from 16 December onwards. The initial reception area was Blackadder Camp at Al Jubayl, Saudi Arabia's principal oil terminus on the Gulf shore. The camp has been described as 'long regimental lines of green Army marquees in the timeless tradition of British Expeditionary Forces'. Not surprisingly, the Commanding Officer's principal aim was to collect the Regiment together and start training for war under desert conditions, but it was not until 1 January that the Regiment had sufficient manpower in Theatre to move out into the desert. The tracked vehicles went

Scimitars leaving Harewood Barracks Herford for embarkation to the Gulf

on low-loaders, but the move was somewhat drawn-out, if only for the lack of maps. Wheeled vehicles had to follow the convoy which, it had to be assumed, knew where it was going. 'Just head up the road going north for 100km until you see a guide!' As anyone who has had experience of desert conditions will know, it is all too easy to get lost.

Once the Regiment was concentrated out in the desert north-west of Al Jubayl, the training was fast and furious as 15 January — the date by which the Iraqis had to pull out of Kuwait — came ever closer. It was during this training that the Regiment unfortunately lost its only officer casualty. The Scimitar which Lieutenant Edward Whitehead was commanding plunged into a quarry in the dark, killing him, although Lance Corporals James and Morgan, who made up the crew, escaped without serious injury. Edward Whitehead was a very promising and popular young officer and his untimely loss was deeply regretted by all ranks. His body was flown home for burial in his native Shropshire.

Few in the Regiment had previous experience of desert operations. Although pleasantly warm by day, the nights could be bitterly cold. Many wished they had brought their parkas with them. Also the desert, although flat and featureless, was a mass of tiny ridges created by the wind which blew unceasingly for much of the time. This made for uncomfortable travelling in short wheel-based vehicles like the Scimitars and Land Rovers, although the Challenger tanks flew across the desert, levelling a passage by sheer weight; as did the Warriors, the newly introduced Infantry Fighting Vehicles. The sudden onset of rain also came as a surprise for those who thought the desert was always dry, and there was more than a little rain in the desert during the winter of 1990-91.

After their seizure of Kuwait the Iraqis started to fortify the 150 miles of border between Kuwait and Saudi Arabia. Their defences were of the linear type and protected by sand berms, mines and barbed wire. The formations manning these defences were mostly 2nd Line troops, poorly trained and equipped. They were supported by tanks and artillery held further back. During their war with Iran the Iraqis had become expert in the

Scimitars training prior to the operation

Scimitars preparing to move. *Soldier Magazine*

B Squadron Swingfire anti-tank guided missile

Numerous injections were given to counter the threat from biological and chemical weapons. Major Walker, Regimental Medical Officer and Lieutenant Ghent

Kate Adie, BBC Television correspondent, interviews Lieutenant Peter Clunie

B Squadron collecting prisoners

Iraqi T55 Tank

Photographs by Captain William Fisher

construction of field defences, which the Iranians had tried to overwhelm by sending wave after wave of untrained conscripts against them, only to be shot down in their thousands. Clearly they hoped to repeat these tactics against the Americans and their Allies, but the Allied success in winning the Air War meant the Iraqis were heading for defeat sooner rather than later.

The frontiers of Kuwait, Iraq and Saudi Arabia meet at the eastern point of what is called the Neutral Zone, about 100 miles inland from the Gulf. From there the Kuwait-Iraq border runs north-east along the Wadi al Batin for about 80 miles, until it turns sharply east for another 80 miles before it reaches the Gulf shore. Between the Neutral Zone and the Euphrates, some 150 miles, the terrain is marked on the map as 'Stony Desert'. There is mile after mile of flat, featureless desert where, after the winter rains, Bedouin and shepherd tribes pasture their camels, goats and sheep, until the scanty winter herbage withers in the sun. It is good territory for fighting, if not for much else. Under normal circumstances a wheeled or tracked vehicle can go almost anywhere, but there are traps for the unwary. Here and there are flat plains of clay-like sand that become lakes after rain, and the edges of these *sabkas*, as they are called, will soon bog down a vehicle. The nature of the 'going' also slowed down the CVR(T)s, to the extent that they were often outpaced by the Challengers.

The Iraqi border with Saudi Arabia had been fortified by the erection of great berms of sand, between which mines had been liberally sown. Behind these berms there were widely dispersed defended localities, held with armour, infantry and artillery. Farther back still were the élite formations of the Iraqi army, the Republican Guard divisions, equipped with the latest T72 Soviet tank. From 16 January onwards, however, all the Iraqi defended localities had been subjected to remorseless bombardment by land and air, wrecking their supply and ammunition dumps, and making a nightmare of their logistic support. But the real ace in the Coalition pack was complete air supremacy. The allied troops had little to fear from air attack and could therefore move closed-up when out of range of Iraq's artillery.

On arrival in the Gulf the Regiment was reinforced by A Squadron of the 1st The Queen's Dragoon Guards. It had already been joined by a troop of the 9th/12th Royal Lancers, and some thirty other officers and soldiers from a wide variety of regiments. In addition the Regiment was supported by 49 Battery Royal Artillery and 73 (OP) Battery Royal Artillery. The Regiment was organized into a Regimental Headquarters, Headquarter Squadron and four Sabre Squadrons — A, B, C, and A Squadron QDG. Each of the sabre squadrons had three reconnaissance troops of four Scimitars, a guided weapons troop of four Strikers and a support troop of four Spartans. The total strength was 750 men, 164 armoured vehicles and 92 soft-skinned vehicles.

The Regiment was grouped under the command of the Commander Royal Artillery, Brigadier Ian Durie, with a view to a somewhat unusual, if not novel, role. Instead of employing the Regiment in its conventional role of reconnaissance, General Smith chose to utilize its mobility and excellent communications in order to seek out and destroy enemy positions in depth by directing artillery fire and air attacks on them.

General Schwarzkopf's overall plan for the destruction of the Iraqi forces defending Kuwait was given the codename DESERT STORM. The 18 (US) Corps on the left, together with the French contingent, were to move rapidly to cut the main Baghdad-Basra highway, about two hours drive south of Baghdad itself. This would prevent reinforcement of the southern battlefield. Simultaneously the 1 (US) Infantry Division was to breach the berms and lift the mines to clear a path through the Iraqi lines for the 7 (US) Corps to pass through; 1 (UK) Armoured Division was to secure the right flank of the 7 (US) Corps, whose task was to bring to battle and destroy the Iraqi Republican Guard Divisions north of the Kuwait border. This was to be the point of main effort. The US Marine divisions were to attack up the coast of the Gulf, and the pan-Arab force (mostly Egyptian and Saudi) was to attack the defences to the east of the Wadi al Batin. The British task was to seek out and destroy those Iraqi formations which might

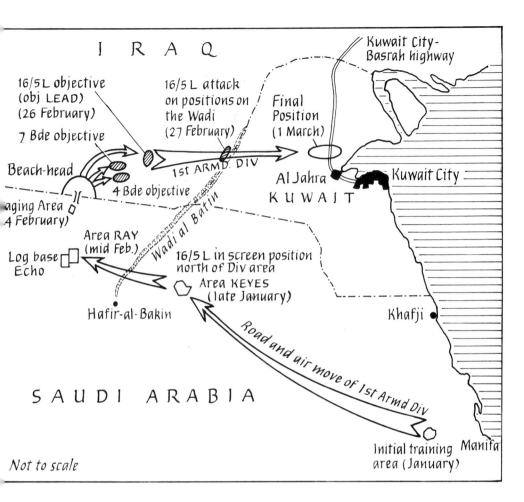

I R A Q

Kuwait City-
Basrah highway

16/5 L objective
(obj LEAD)
(26 February)

7 Bde objective

16/5 L attack
on positions on
the Wadi
(27 February)

Final
Position
(1 March)

Beach-head

1st ARMD. DIV

4 Bde objective

Al Jahra

K U W A I T

Kuwait City

aging Area
4 February)

Area RAY
(mid Feb.)

Log base
Echo

Wadi al Batin

16/5 L in screen position
north of Div area

Area KEYES
(late January)

Hafir-al-Bakin

Road and air move of 1st Armd Div

Khafji

S A U D I   A R A B I A

Not to scale

Initial training
area (January)

Manifa

threaten the right flank of the advancing 7 (US) Corps. It was
reckoned that these might be the Iraqi 12 and 17 Armoured
Divisions and the Tawalkana Republican Guard Division. The
start line on the Iraqi side of the frontier defences was
codenamed NEW JERSEY and the initial objective was the line
SMASH, some 100 km into Iraq from the breach. It was
reckoned this might take ten days of intensive fighting; in the
event SMASH was reached in less than 72 hours.

During the latter part of January, 1991, the 1st Armoured
Division moved by road and air from its initial training area
north-west of Al Jubayl to area KEYES, 50 miles south of the
point at which the Kuwait, Iraq and Saudi borders meet. Here

the Regiment provided a protective screen for the Division and took part in rehearsals for the breach crossing operation. The Division then moved farther west to area RAY which was some 140 miles from the coast and about 40 miles south of the Iraqi border. On 23 February, on the eve of the ground war, the artillery was bombarding enemy positions, under the protection of A Squadron QDG; the guns had begun firing on 19 February. On the morning of 24 February Lieutenant-Colonel Scott was summoned to receive orders at Divisional headquarters, while the Second-in-Command (Major Alick Finlayson) moved with the Step-Up headquarters to join with the 1st (US) Infantry Brigade headquarters to coordinate the Regiment's passage through the breached positions.

General Smith's plan was for 7th Armoured Brigade to lead the breakout, with 4th Armoured Brigade to follow. They were to attack sequentially a number of identified Iraqi positions, as shown in the attached diagram. Each of the 'goose eggs' in the diagram marked COPPER, ZINC etc represents an Iraqi defended locality of brigade or battalion group in strength, supported by tanks and artillery. The task given the Regiment was to 'fix', and then 'engage' with aircraft, Multi Launch Rocket Systems (MLRS) and artillery fire Iraqi depth positions. The two armoured brigades following behind were to attack and destroy identified Iraqi positions in a series of 'bite-sized chunks'. The Regiment, by attacking the enemy in depth, would prevent him from reinforcing or counter-attacking in support of the objectives being attacked by the brigades. This was certainly an unusual role for an armoured reconnaissance regiment with its lightly armoured vehicles; the more usual role being 'to see without being seen', and although there will be occasions when it will have to fight for information, this will happen only infrequently. More often in the advance it will be employed guarding a flank or in mobile reserve.

In fact the Regiment did achieve what had been required of it, principally with the aid of MLRS, which turned out to be a battle-winning weapon. But the narrow frontage and limited real estate for deployment of four reconnaissance squadrons, coupled with the fact that the main battle tanks following up behind found enemy resistance light and the 'going' easier for them than for the CVR(T)s, meant that sooner rather than later the armoured brigades would catch up with the Regiment, thus restricting even further the Regiment's room for manoeuvre. Moreover, the Regiment's role meant almost certain confrontation with the much heavier Iraqi armour until the Challenger could deliver the coup de grâce. However, in the event, the sheer speed and ferocity of the Allied attack stunned the Iraqis and their armour displayed little will to fight.

Since the Regiment was to move to a staging area at 1500 hours, where it would be joined by A Squadron QDG, Lieutenant-Colonel Scott had to act quickly. The Echelon moved off to join up with the Artillery Group echelon, under whose direction it would now be operating, while RHQ and the sabre squadrons got going. The staging area was about 20 miles south of the border and the Regiment was complete there by 1800 hours on 24 February, where the final preparations were made. The absence of any enemy Air made it possible to move under less dispersed conditions than would have been possible in, say, Italy during the Second World War, but there was of course always the threat from the enemy artillery. 1 (US) Infantry Division had made excellent progress in breaching the enemy's defences, but in order to avoid confusion halted for the night of 24/25 February.

When dawn came on 25 February the Regiment was raring to go but was not called forward until 1000 hours. This was frustrating because it was hoped to lead the two Armoured Brigades into Iraq, but in the event 7th Armoured Brigade moved first, but on a different axis. There was inevitably some confusion with so many armoured vehicles to be passed through the minefield, which was carefully controlled by the Americans, and it was not until the Regiment reached its Forming Up Point (FUP) at VALLEY FORGE at 1400 hours that its breakout became imminent. After pausing for about 20 minutes at VALLEY FORGE, the Regiment was called forward, crossing the Start Line (NEW JERSEY) at 1500 hours. It was now in enemy-held territory, with 7th Armoured Brigade on its inner flank, but heading for different objectives. The Regiment had

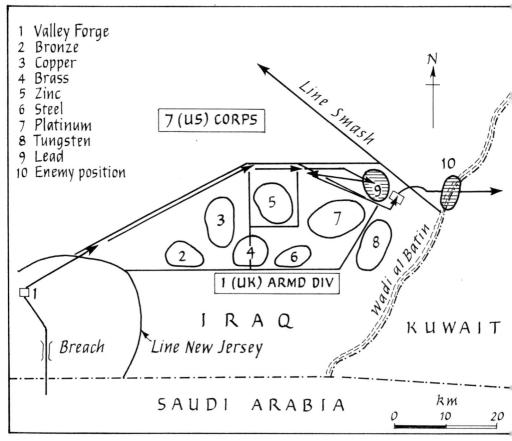

1 Valley Forge
2 Bronze
3 Copper
4 Brass
5 Zinc
6 Steel
7 Platinum
8 Tungsten
9 Lead
10 Enemy position

7 (US) CORPS

1 (UK) ARMD DIV

N

Line Smash

Wadi al Batin

I R A Q

Breach

Line New Jersey

K U W A I T

S A U D I   A R A B I A

km
0   10   20

16/5L Movements – 25–27 February, 1991

been ordered to bypass COPPER and head for ZINC.

There had just been time at the FUP for the Regiment to shake out into battle formation and it set off at a great pace, led by B Squadron (Major Richard Quicke of the 13th/18th Royal Hussars); followed by C Squadron (Major Roly Rickcord), A Squadron QDG (Major Hamish Macdonald) and A Squadron 16/5L (Major Charlie Grant). Major Mark Ridley was the Operations Officer and Captain Nick Wills was the Adjutant, both with RHQ. The Echelon (with the Artillery) was commanded by Headquarter Squadron Leader, Major Bill Cook.

Following immediately behind the rearmost squadron was a small group under the MTO, Captain Tony Willmore. This consisted of two fuel vehicles, an armoured collecting section with a Doctor, five sections from 73 (OP) Battery in FV 432s and the EME (Captain Billy Joels) with two fitters vehicles.

73 (OP) Battery was attached to the Regiment in order to assist with directing artillery and air on to depth targets. Were the Regiment to be required to bypass an enemy position, it could drop off one or more of the battery's sections to engage the enemy with artillery fire if necessary. The Battery also provided a Fire Support Coordination Cell, an Air Coordination Cell and two Forward Air Controllers (FACs). As such it was to render invaluable service.

Towards nightfall the weather began to worsen and it started to rain. What is more, the CVR(T)s with their short tracks were finding that the corrugated surface of the desert made for very uncomfortable travelling; so much so that both Challenger and Warrior were able to move faster across the desert. Although there should have been a full moon, it was obvious that there would be little visibility that night; and to add to the problem there was a driving wind which lashed the rain against the vehicles.

There was a short halt just before dark in order to deconflict with 7th Armoured Brigade and give any stragglers the opportunity to catch up. It had been discovered that the Regiment's task, which was to screen ZINC from the north and engage it with artillery fire, clashed with 7th Armoured Brigade's, who were moving to attack ZINC from the north. This led to considerable confusion, B Squadron narrowly escaping being shot up by the Royal Scots Dragoon Guards who had attacked COPPER, while the Staffords regimental group heading north drove straight through the Regiment. Luckily communication and liaison with the 7th Armoured Brigade was good and a firefight with our 'own side' was averted.

Just before midnight the Regiment moved into a tight assembly area to the north of ZINC in an attempt to keep out of 7th Brigade's way. Soon afterwards they heard and saw the devastating weight of fire being delivered against ZINC. The

Iraqis were later to claim that the artillery fire directed against their tanks and artillery pieces was much more devastating than the bombing to which they had been subjected from the air. The MLRS were awesomely effective, scattering their bomblets over a wide area.

Although by this time the squadrons had covered the best part of 50 miles in extremely adverse weather conditions, there had been no contact with the enemy, apart from seeing a dozen or so Iraqis seeking to surrender as the squadrons hurtled past. Soon after midnight the Regiment's objective was changed and it was ordered to screen LEAD and attack it with artillery and air. It was 0300 hours on 26 February when the Regiment moved forward and took up position round LEAD. C Squadron and 73 (OP) Battery faced north, from which the most likely counter-attack would come. B Squadron and A Squadron QDG covered the western edge of LEAD, and A Squadron 16/5L faced south to guard the right flank. The MTO and his group were left behind in the overnight assembly area, and RHQ and the Regimental Aid Post took up a central position. Then everyone waited for the dawn.

As soon as it was light enough to see, B Squadron and A Squadron QDG set about identifying enemy positions within LEAD. MLRS had moved up to support the Regiment and the first artillery fire was called down at 0647 hours. MLRS continued to support the Regiment 'with devastating accuracy and effect' throughout the morning. It was clear that LEAD contained a substantial Iraqi force, which included tanks and APCs. It was almost exactly 46 years since the Regiment was last engaged in full-scale battle − in Italy on 23 April, 1945, not far from the banks of the River Po. Not one of the 'Scarlet Lancers', from the Commanding Officer to latest-joined Trooper who were fighting in Iraq on the morning of 26 February, had been born when the Regiment last went into action.

A pair of US 'Tank-Buster' A10 aircraft were directed on to the Iraqi positions, destroying six tanks and a number of APCs. Several Iraqi tanks, dislodged from their dug-in positions, began to head west in the direction of B Squadron, who engaged them with their 30 mm cannon. It was, however, the Swingfire guided missiles of Lieutenant Philip Lepp and Corporal Carl Radford which knocked out three tanks and an APC. Meanwhile 7th Armoured Brigade was attacking ZINC with great effect, prior to switching their attention to PLATINUM. By blocking off LEAD the Regiment was effectively preventing any reinforcement of PLATINUM, but it did mean that enemy tanks seeking to escape north would as likely as not run head on into the Regiment's much more lightly armoured vehicles. This kept all squadrons on their toes as the two armoured brigades to their south tore the Iraqi positions apart.

To add to the confusion which is always inseparable from battle, the weather continued to play tricks. The rain had been followed by a driving wind that whipped the desert into a sandstorm. This proved to be almost impenetrable to thermal imaging sights and enormously complicated the identification of targets. The effects of the bombardment of LEAD were akin to stirring up a hornets' nest. From 0800 hours until midday all four sabre squadrons were busily engaged on no less than three fronts. An enemy attempt to reinforce LEAD from the north was stopped by C Squadron, who knocked out the enemy's two lead tanks with Swingfire guided missiles at a range of more than 3,000 metres. B Squadron had its hands full directing artillery fire and engaging Iraqi tanks and APCs as they emerged through the clouds of sand.

A Squadron QDG was equally busy directing artillery fire and controlling aircraft strikes on LEAD. They destroyed both tanks and APCs. A Squadron 16/5L, who were covering the southern flank, found themselves right in the path of enemy tanks fleeing from 7th Armoured Brigade. Lieutenant Jamie Horton's Troop stopped a T62 after hitting it more than 30 times with 30 mm cannon fire. A Striker, which had managed to acquire two prisoners, scored a glancing hit on the turret of an advancing tank. The Iraqis standing behind the vehicle were engulfed in smoke and flame when the missile was launched. Regardless of their smouldering trousers, however, they advised the crew to aim off a little to the left!

As the sandstorm gathered strength, it was like driving through a drifting fog. Vehicles became indistinguishable at one moment, and in the next loomed up as dark shapes – friend or foe? Then it would suddenly clear, sometimes to reveal an Iraqi tank bearing down out of the curtain of sand. Several 'rogue' tanks were roaming the battlefield, presumably seeking to escape north, but representing a threat to whoever got in their way. The wind howled like a banshee, and the electricity in the air seriously interfered with radio communications. RHQ lost contact with Division for a time but fortunately was able to relay messages via 7th Armoured Brigade. RHQ was particularly vulnerable to roving Iraqi tanks, but the Colonel decided to remain put and hoped to avoid detection. To add to the problems, there were individual Iraqi soldiers wandering about and seeking to surrender. Four who were taken by B Squadron turned out to be doctors, the senior of them a Lieutenant-Colonel. By 1030 hours it was clear that several Iraqi tanks had managed to get behind the squadron, increasing the threat from every direction. Meanwhile the sand continued to blow. The 'Fog of War' as it might be described.

Lieutenant Morley's Scimitar had been holed by heavy machine-gun fire from a T59 tank which had set on fire some of the ammunition stored in the turret. Morley and his crew hurriedly baled out to find what cover they could. When no explosion followed, they were again shot up when they tried to reoccupy the vehicle. They had to wait until the T59 had moved off into the gloom before returning to their vehicle, Trooper Wakelam, the driver, with a broken ankle. The crew of a disabled Scimitar, which was on tow, were also engaged by a T59. They replied with automatic fire and the Iraqi sheered off in search of easier prey.

This it found in the shape of a couple of fuel and ammunition vehicles crewed by REME. They were M548 tracked vehicles provided by the Americans and belonging to C Squadron. SSM Scott was leading them in a Ferret. One of the M548s broke down, whereupon the other M548, commanded by Sergeant Michael Dowling, moved in to pick up the crew, the Iraqi having temporarily lost sight of them in the sandstorm. However, the visibility improved and the Iraqi returned to the chase, firing with its heavy calibre machine-gun. Lance-Corporal

Francis Evans, who was sitting between the Driver and Sergeant Dowling, was hit in the chest, dying instantly. Sergeant Dowling, who was trying to distract the enemy's tank aim by leaning out of the cab and firing his rifle at it, was also hit and killed. Meanwhile the chase continued until the Iraqi tired of it and moved off elsewhere. SSM Scott, who was trying frantically to lead the M548 to safety, was so thrown about in the short wheeled-base Ferret that he had to be evacuated as a casualty with a badly strained back after the chase was over. The bodies of Sergeant Dowling and Lance-Corporal Evans, both of them REME, were flown home later and buried with military honours. Sergeant Dowling was subsequently awarded a posthumous MM for his brave conduct during this action.

To add to the confusion it was reported that a column of enemy tanks was approaching from the west and the Guided Weapons Troop from A Squadron was sent to head them off. This succeeded in diverting them in the direction of the Queen's Royal Irish Hussars who obligingly destroyed the lot. It was now midday on 26 February and no more concentrations of vehicles could be seen on LEAD. The principal threat came from Iraqi tanks, both collectively and individually, seeking to escape north through the area then occupied by the Regiment. Moreover, 7th Armoured Brigade had been tasked to attack LEAD and the Regiment's dispositions would get in its way. Lieutenant-Colonel Scott therefore ordered the squadrons to move sideways to the concentration area where they had spent the previous night; there they could replenish the Scimitars and Strikers. Since early morning the Regiment, with its much lighter vehicles, had been engaged with the equivalent of an Iraqi armoured brigade, and had rendered it non-effective. For the Regiment's first baptism of fire since the end of the war in Italy in 1945, everything had gone surprisingly well.

From about 1430 hours until nightfall, the Regiment was busy with replenishment from Captain Willmore's echelon, the repair of vehicles by the fitters and the handing over to the nearest collection point of a mixed bunch of prisoners. All vehicles beyond immediate repair were grouped together for handing over by the EME to the nearest equipment collection point. Meanwhile over the radio there came a constant succession of orders and counter-orders. It was clear that the Divisional Commander was receiving changing directions from 7 (US) Corps as the American advance on the left proceeded with unexpected speed and success. This did not make it easy for the Commanding Officer to plan the next move. Originally the task had been to lead the Division in order to find the next Iraqi defended locality and engage it from a flank until the two following armoured brigades could come up to destroy it, but there was no indication of the new direction the Regiment was to take. By 0230 hours 27 February, however, the Regiment was in position for the further advance, but did not expect to cross the Start Line until after 0600 hours. This almost gave the Colonel the chance to snatch a little much-needed sleep, but it was not to be. He lay down expecting the advance to continue up the Wadi al Batin north-eastwards, but minutes later at 0320 hours the orders changed, giving a different direction of advance and a different Start Line. This involved hasty redeployment against the clock by weary crews, and even wearier squadron leaders. The axis of advance had been changed from north-eastwards to eastwards directly into Kuwait, across terrain for which the Regiment held no maps. Captain Wills, the Adjutant, was dispatched to HQ 7th Armoured Brigade to beg, borrow or steal some, while at the same time A Squadron QDG was detached and placed under command of 7th Armoured Brigade.

Despite the short notice, and the inevitable confusion that results from changes of plan in the middle of a battle, the Regiment crossed the Start Line at dawn on 27 February with A Squadron leading on the left (north), and B Squadron on the right (south). C Squadron was in reserve. The mission was to clear the enemy in the area of the Kuwait/Iraq border to the Regiment's front, in order to clear a passage for 7th Armoured Brigade's drive towards Kuwait City. The advance was not as rapid as previously. Both leading squadrons came up against a very strong Iraqi position straddling the Wadi al Batin, which contained dug-in infantry, tanks and artillery. A10 aircraft and MLRS fire were directed against it with great success but it took

Lieutenant-Colonel Philip Scott at Regimental Headquarters

some time to shift the Iraqis. By then 7th Armoured Brigade, which was motoring at a great rate across the broken terrain, had caught up with the Regiment. As the brigade thundered into Kuwait, the Regiment pulled aside to clear the way.

Lieutenant-Colonel Scott then concentrated the Regiment about six miles inside Kuwait. It was now midday. When orders came from Division, they were hardly as expected. Instead of carrying on into Kuwait, the Regiment was ordered to move back into Iraq, in order to provide security for the many logistic units following behind the two armoured brigades. There were of course large numbers of Iraqis milling around in the desert trying to surrender. The Regiment moved off at about 1800 hours just as darkness fell. It turned out to be a 50-mile drive in pitch darkness by drivers already exhausted from three days of battle. Every crewman was equally tired. The Divisional Rear Area was reached by the small hours, when as many men as could tried to snatch some sleep.

They awoke to hear the glad news at 0800 hours on 28 February of a cease fire. There still remained the danger that news of it had not reached all the enemy, some of whom might continue to fight. Judging from the remarks of some of the prisoners, anything was possible with an enemy who appeared to be expecting attack not from the British or Americans, but from Israeli lorried infantry! Clearly in the Iraqi Army the passage of information downwards is just as bad as the passage of information upwards.

A Squadron was detached to provide protection for the Prisoner of War Guard Force. The remainder of the Regiment was warned to escort the Division's logistic 'tail' into Kuwait. This entailed a move of nearly 100 miles across the desert. The convoy moved off around 1800 hours and we are told that 'the move that night was unspeakably dreadful. Visibility was still not good, the speed of the convoy was tortuously slow, and the terrain was extremely bumpy'. There was also the problem of dust that added to the difficulties. The convoy travelling with headlights switched on resembled a vast traffic jam that had escaped the M25. For B and C Squadrons travelling to the north and south of the convoy, the journey was complicated by the litter of the battlefield, abandoned Iraqi positions, unexploded bombs and bomblets, minefields, and hundreds of Iraqis trying to surrender. Shortly after first light on 1 March the convoy was delivered to its new position 40 miles north-west of Kuwait City. The Regiment's B Echelon, from which it had been separated for nearly six days, caught up with it again in the afternoon of 1 March. A Squadron rejoined two days later.

In words that bear repetition it has been written that 'The Regiment had been on the go almost continuously for five days and had taken part in two major actions. We had operated ahead of the Division against enemy armour in old, lightly armed and lightly armoured vehicles. We had covered about 300 miles, many of them in conditions of appalling visibility. We had achieved the tasks given to us, suffering light but nevertheless tragic casualties. The Regiment can feel justifiably proud of the way it operated as a team and successfully overcame the many difficulties it encountered.'

'The men performed brilliantly, with courage and stamina,' said their Commanding Officer. They showed themselves to be true professionals. The Regiment was the first British unit to enter Iraq, and the first to enter Kuwait. It was the only British regiment to call for and direct artillery and air support which had such a devastating effect on the enemy. In the last battle the Regiment is likely to fight under its present title, the 'Scarlet Lancers' certainly sounded off with a flourish. 'Gentleman Johnny' Burgoyne would have been proud of them.

During the battle they knocked out with cannon fire and Swingfire missiles two T62 and seven T55 main battle tanks, as well as 15 armoured personnel carriers. This had involved an expenditure of 248 rounds of 30mm, and 26 guided missiles. Together with the Staffords, our County infantry regiment, they were the only two regiments to have soldiers killed by enemy fire; sadly there were casualties suffered by other regiments from friendly fire. Of the soldiers who served with the Regiment in Iraq, roughly 35 percent came from Staffordshire and the West Midlands. Everyone on the regimental list fit to do so took part in the action, including the Band who served as Regimental Medical Assistants, and they were in the thick of it.

Quite apart from the remarkably short duration of the land battle and the evident unwillingness of the ordinary Iraqi soldier to fight, the land war in the Gulf possessed some special features which are worthy of comment in a regimental history such as this. In the first place the British soldier showed himself to be as adaptable and as good-humoured as his forefathers had been before him. He took the heat, the sand-laden winds, the bitter cold and the rain in his stride. Even the absence of beer, and the presence of vipers, camel spiders and scorpions, failed to shake him. Few of them had been in action before but their attitude towards it was well summed-up by a letter from one of the junior subalterns to his parents, when he wrote, 'It's very exciting in a scary sort of way'. Their pride in the Regiment and their own professionalism was uppermost in their minds. As they waited during the tense hours for the order to break out into Iraq, it was probably not patriotism, nor fear of casualties, which dominated their minds. It was the desire not to let down the Regiment that most filled their thoughts – and none of them did.

They did of course possess certain advantages over those who had fought in the Regiment before them. The first was, of course, complete command of the Air, an advantage of incomparable value. Another was the ineptitude of the enemy. Then, compared with those of us who had struggled with the difficulties of navigation in the desert in previous campaigns – the sun compass and the like – the Global Positioning System (GPS) satellite navigation system was issued down to Troop level. With its aid the Regiment could navigate in the otherwise featureless desert with unerring accuracy and at speed. It proved its worth time and time again. It was no longer easy to get lost! And finally there were the MRLS batteries of the Royal Artillery whose accuracy and devastating effect wreaked havoc on Iraqi morale. There were those among them who said they preferred bombardment from the USAF B-22s to a salvo from the MRLS.

It was also the case that this was one of the first wars to be fought by the British Army under the close scrutiny of the television cameras and commentators. As the long drawn-out negotiation to make Saddam Hussein see sense filled the television screens, the media correspondents indulged in a great deal of conjecture, dwelling overmuch perhaps on the risk of casualties, to the extent that 'body bags' became part and parcel of the English language. If they did not frighten the enemy, they certainly frightened some of their viewers. They were the television pundits sitting at home, interpreting the news as they saw fit, and frequently wrong. The television correspondents in the field were, however, a different breed altogether. Visits by them to the Regiment were always welcomed, and among them perhaps the indomitable Kate Adie ranked highest for her courage and sympathetic approach.

Finally, there were the families. Previously, when the Regiment had gone to the wars, the families had been scattered

Welcome home from the families

far and wide throughout the nation. But in this instance they were mostly all together at Herford in Germany, far from their own parents and family friends. No account of the Regiment in the Gulf War would be complete without a reference to the sterling work of Captain Richard Bennison, the Families Officer, and Mrs. Diana Scott, the 'Colonel's Lady'. On their shoulders rested the responsibility of looking after the anxious families left on their own in Germany. The fact that morale remained so high says a very great deal for the success of their efforts. It was an unusual feature of the war that before and after the land battle soldiers were able occasionally to telephone their

255

families, and the Forces air letter cards — 'blueys' as they came to be known — were eagerly awaited as much by the families at home as they were by the soldiers in the field.

It had been a 'Famous Victory', and probably the last one the 16th/5th Lancers would fight under their present title. The part they had played in it was marked by the subsequent awards of one M.M. (posthumous); two B.E.M.s; five Mentions in Dispatches; two C-in-C's Commendations; one US Army Bronze Medal and one US Army Commendation. It must be remembered, however, that it was a *Coalition* victory, with the Americans very firmly in the driving seat. All those who fought alongside American soldiers were full of praise for them. As one of the Regiment described them: 'They were different from the soldiers we had read about in Vietnam. These soldiers were "lean, fit and mean", very professional in the way they went about their business, and we were glad they were on *our* side!' Due tribute must also be paid to the Airmen who fought the Air War which gave us complete air supremacy, and to the Navies who dominated the Gulf. From beginning to end it had been a true team effort.

After victory had been won, the Regiment did not linger long in the Gulf but was back in BAOR by the end of March to a magnificent reception organized by Captain Bennison and his rear party. All things considered it had been quite an exciting three months.

Major-General Rupert Smith, GOC 1st (UK) Armoured Division, has had the following to say about the Regiment's part in the battle:

'Although the Divisional attack lasted only a little over 24 hours the Regiment played, literally, a leading role. They were well out in front of the Division and their aggressive and accurate use of the firepower available defeated the enemy's attempts to move to reinforce his defence.'

EPILOGUE

IT would be wrong to pretend that everything has gone well for the Regiment throughout the three hundred years of its existence. A glance at the chequered history of the Royal Irish Dragoons would be sufficient to dispel any such notion. After both the two world wars of this century the Regiment needed time to regain its former *élan*, largely owing to the departure of most of the officers and NCOs who had served in it during the wars. Nor has the fault always lain with the Regiment. Between the two world wars, when the Army was starved of money and recruiting was difficult, far too much of the Army's time was devoted to sport, and to money-raising events such as Pageants, Tattoos etc. One consequence of this in 1926 was that the 16th/5th Lancers were short overall of 76 soldiers (almost a squadron) and of the remainder no fewer than 116 were required daily for garrison fatigues in Tidworth. This left only eight men per squadron to groom upwards of 100 horses per squadron. It is small wonder that the Inspector-General of Cavalry expressed his displeasure at their appearance and condition.

What makes a 'Good' regiment? All regiments consider themselves to be equally good, but as in the case of George Orwell's pigs, some are more equally good than others. Money used to have something to do with it. Expensive regiments attracted wealthy officers who could afford to set the pace at a time when uniforms and drill were the main criteria for military efficiency. They became known as 'crack' regiments, an expression much-favoured and often over-used by journalists when writing about the Army. But since this century began, war has become much more complicated and uniform much more utilitarian. Neither money nor breeding will guarantee a pass at the Regular Commissions Board, nor out of Sandhurst, nor into any particular regiment. The days of purchasing commissions ended more than 100 years ago, and the requirement in certain regiments that their officers should possess some kind of private income is as dead as the dodo. What is more, the days of the amateur, when prowess in the hunting field or on the polo ground could make up for deficiencies of a military kind, have long since ended.

Today regiments are judged by their efficiency and no longer by their appearance on parade or the depth of their officers' purses. Today's soldier has to feel that he is doing a worthwhile job. He still has pride in his regiment, which for most soldiers is probably the limit of their army horizon, but he has no use for the officer or NCO who is not pulling his weight. Soldiers nowadays have often to carry out their duties in front of television cameras, and are required to display great restraint in the face of intense and unmerited provocation, as in Northern Ireland. They have to operate immensely complicated equipment. They may not be quite so physically tough as their forbears because they have grown up in a different environment, but they are just as capable of feats of great gallantry. Such as Sergeant Michael Dowling who was killed when attempting to scare off an Iraqi main battle tank by firing at it with his rifle, for which self-sacrificing gallantry he was awarded the Military Medal posthumously.

The modern British soldier is probably more ambitious and less content to remain in the same old groove as so many of his predecessors were. If soldiers know their job thoroughly and take a pride in doing it; if they respect their officers as equally competent professionals, and are in their turn respected; if they believe that they and their families are getting a square deal from the Army and the Nation — then they will be 'Good' soldiers, and the regiment in which they serve will be a 'Good' regiment.

It has been said that the regiments of the British Army are like families; and like most families they have their good times and their bad. This family-likeness has been strengthened by our link with Staffordshire which began in 1958, since when brother has followed brother, and friend has followed friend, from all over Staffordshire and the West Midlands into the 'Scarlet Lancers'. The same has been true of the officers. When the regiment was serving in Aden in 1964 there were no less than five sons or grandsons of previous commanding officers serving as subalterns with the Regiment.

When John Burgoyne first raised the 16th Light Dragoons in 1759, he was far in advance of his time when he encouraged his officers to cultivate a friendly relationship with their soldiers, who were for the most part regarded as mere cannon fodder. This has been a characteristic of the Regiment down the years, as can be evidenced by the career of Trooper William Robertson who rose through all the ranks, non-commissioned and commissioned, to end as a Field-Marshal and head of the British Army. His Troop-Sergeant-Major's scarlet stable jacket hangs today in the Warrant Officers' & Sergeants' Mess as an example of what can be achieved by any 'Scarlet Lancer' with the courage, determination and will to succeed; but he has as yet to be equalled!

In the British Army cavalry regiments have long been considered as 'stuck-up' and 'stand-offish'. The Regiment has always sought to avoid such a label being attached to it. 'We got on well with all the other regiments,' wrote John Luard of Bhurtpore in 1825. 'The 16th Lancers never walked near our cooking places nor spat on our food,' wrote Sita Ram, a high-caste Hindu, when comparing the 16th Lancers favourably with the other European regiments in the First Afghan War in 1840. General Gough, writing of the 16th in India in 1897, has said, 'There was no snobbery in the 16th. We got on well with all regiments alike. We had a good opinion of ourselves but took good care not to show it.' It is a tradition well worth preserving.

This has been the story of two distinguished cavalry regiments of ancient lineage who came together in 1922 to create a third equally distinguished regiment, the 16th/5th The Queen's Royal Lancers, which has showed itself worthy of its two distinguished forbears in both peace and war in the seventy years that have followed. The story ends on the eve of yet another change in the Regiment's fortunes when, as a consequence of the planned reduction in the Royal Armoured Corps, it is to be amalgamated with another distinguished

Cavalry regiment of equally great traditions, the 17th/21st Lancers. Raised in the same year as the 16th, 1759, the 17th Light Dragoons charged at Balaclava in 1854, and again at Ulundi against the Zulus in 1879. In 1922 the Regiment was amalgamated with the 21st Lancers, whose charge at Omdurman against the Dervishes in 1898 is chiefly memorable for the fact that Lieutenant Winston Churchill, attached from the 4th Hussars, charged with the 21st Lancers that day.

Together, in October, 1942, the 16th/5th Lancers and the 17th/21st Lancers went to North Africa in the 26th Armoured Brigade, took part in the triumphant capture of Tunis, and went on together to fight their way up the length of Italy until the war ended on the banks of the River Po in April, 1945. Both are regiments of similar character and record, each proud of their own traditions, but equally respecting the other's. If there has to be a match, this one could hardly be bettered, and everyone, both serving and retired, is determined to make a success of it.

Naturally there will be regrets when two proud regimental titles disappear from the *Army List*, to be replaced by another which, in the fullness of time, will evoke the same sentimental attachment and pride. From their first raising the 16th have always been a 'Queen's' regiment, a connection which has been immeasurably strengthened since Queen Elizabeth II ascended the Throne and conferred on the Regiment the title of *The Queen's Royal Lancers*. Whatever the future may betide, it is very much to be hoped that the Regiment's close connection with The Queen will be preserved in the future. It is a connection much honoured and prized by all who have served in the 'Scarlet Lancers'.

GOD SAVE THE QUEEN.

APPENDIX

Roll of Colonels-in-Chief, Colonels, & Commanding Officers
1925-1992
Colonels-in-Chief

H.M. King Alfonso XIII of Spain	1905-1941
H.M. Queen Elizabeth II	1947-

Colonels

Lieutenant-General Sir James Babington, KCB, KCMG	
	1909-1936
Field-Marshal The Viscount Allenby, GCB, GCMG	1912-1936
General Sir Hubert Gough, GCB, GCMG, KCVO	1936-1943
Colonel H.C.L.Howard, CB, CMG, DSO	1943-1950
Brigadier P.E.Bowden-Smith, CBE	1950-1959
Colonel D.D.P.Smyly, DSO	1959-1969
Colonel A.S.Bullivant, MBE	1969-1975
Major-General J.D.Lunt, CBE	1975-1980
Colonel H.A.G.Brooke, MC	1980-1985
Brigadier J.L.Pownall, OBE	1985-1990
Major-General A.W.Dennis, CB, OBE	1990-

Commanding Officers (since 1945)

Lieutenant-Colonel D.D.P.Smyly, DSO	1944-1947
Lieutenant-Colonel K.C.C.Smith	1947
Lieutenant-Colonel G.H.Illingworth, MBE	1947-1948
Lieutenant-Colonel T.C.Williamson, DSO	1948-1949
Lieutenant-Colonel D.R.B.Kaye, DSO	1949-1951
Lieutenant-Colonel J.R.Cleghorn, DSO	1951-1954
Lieutenant-Colonel A.S.Bullivant, MBE	1954-1957
Lieutenant-Colonel J.D.Lunt, OBE	1957-1959
Lieutenant-Colonel R.A.Simpson, OBE	1959-1961
Lieutenant-Colonel P.C.Bull	1961-1964
Lieutenant-Colonel P.J.Holland, MC	1964-1966
Lieutenant-Colonel H.A.G.Brooke, MC	1966-1969
Lieutenant-Colonel J.L.Pownall, OBE	1969-1971
Lieutenant-Colonel A.W.Dennis, OBE	1971-1974
Lieutenant-Colonel R.Q.M.Morris, OBE	1974-1976
Lieutenant-Colonel The Hon. N.C.Vivian	1976-1979
Lieutenant-Colonel C.J.Radford, MC	1979-1983
Lieutenant-Colonel J.A.Wright, MBE	1983-1985

Lieutenant-Colonel D.M.O'Callaghan	1985-1986
Lieutenant-Colonel M.F.C.Radford	1986-1989
Lieutenant-Colonel P.E.Scott	1989-1991
Lieutenant-Colonel P.M.Campbell	1991-

BATTLE HONOURS

At the time of their amalgamation in 1922 the Regiment possessed more Battle Honours than any other cavalry regiment of the line.

16th The Queen's Lancers

Beaumont, Willems, Talavera, Fuentes d'Onor, Salamanca, Vittoria, Nive, Peninsula, Waterloo, Bhurtpore, Ghuznee 1839, Afghanistan 1839, Maharajpore, Aliwal, Sobraon, Relief of Kimberley, Paardeberg, South Africa 1900-02, Mons, Le Cateau, Retreat from Mons, Marne 1914, Aisne 1914, Messines 1914, Ypres 1914, '15, Bellewaarde, Cambrai 1917, Somme 1918, St Quentin, Pursuit to Mons.

5th Royal Irish Lancers

Blenheim, Ramillies, Oudenarde, Malplaquet, Suakin 1885, Defence of Ladysmith, South Africa 1899-1902, Mons, Le Cateau, Retreat from Mons, Marne 1914, Aisne 1914, Messines 1914, Ypres 1914, '15, Bellewaarde, Arras 1917, Cambrai 1917, Somme 1918, St Quentin, Pursuit to Mons.

16th/5th The Queen's Royal Lancers

Fondouk, Bordj, Djebel Kournine, Tunis, North Africa 1942-43, Casino II, Liri Valley, Advance to Florence, Argenta Gap, Italy 1944-45.

VICTORIA CROSSES AWARDED TO THE REGIMENT

The Tirah Expedition 1897

Lieutenant The Viscount Fincastle (later the Earl of Dunmore), 16th The Queen's Lancers

South Africa 1901

Lieutenant F.B.Dugdale, 5th Royal Irish Lancers

France 1917

Private George Clare, 5th Royal Irish Lancers

Index

References to illustrations indicated thus: 102*

Index

Index